Rational Design of Next-generation Nanomaterials and Nanodevices for Water Applications

Rational Design of Next-generation Nanomaterials and Nanodevices for Water Applications

Peng Wang

IWA PUBLISHING

Published by **IWA Publishing**
Alliance House
12 Caxton Street
London SW1H 0QS, UK
Telephone: +44 (0)20 7654 5500
Fax: +44 (0)20 7654 5555
Email: publications@iwap.co.uk
Web: www.iwapublishing.com

First published 2016

British Library Cataloguing in Publication Data
A CIP catalogue record for this book is available from the British Library

ISBN: 9781780406855 (Paperback)
ISBN: 9781780406862 (eBook)

Contents

Editor and contributors

Aaron D. Wilson
Idaho National Laboratory, USA.
aaron.wilson@inl.gov

Arturo A. Keller
Bren School of Environmental Science and Management, University of California at Santa Barbara, CA, USA 93106.
keller@bren.ucsb.edu

Brian J. Starr
Department of Civil and Environmental Engineering, Michigan State University, East Lansing 48824 USA.
tarabara@egr.msu.edu

Chengzhong Yu
Australian Institute for Bioengineering and Nanotechnology (AIBN), Corner College and Cooper Rds (Bldg 75), The University of Queensland, Brisbane Qld 4072, Australia.
c.yu@uq.edu.au

Christopher A. Crock
Department of Civil and Environmental Engineering, Michigan State University, East Lansing 48824 USA.
tarabara@egr.msu.edu

Jie Yang
Australian Institute for Bioengineering and Nanotechnology (AIBN), Corner College and Cooper Rds (Bldg 75), The University of Queensland, Brisbane Qld 4072, Australia.
c.yu@uq.edu.au

Lianbin Zhang
Biological and Environmental Sciences and Engineering Division, Water Desalination and Reuse Center (WDRC), King Abdullah University of Science and Technology.
peng.wang@kaust.edu.sa

Muhammad Mustafa Hussain
King Abdullah University of Science and Technology, Saudi Arabia.
muhammadmustafa.hussain@kaust.edu.sa

Peng Wang
Biological and Environmental Sciences and Engineering Division, Water Desalination and Reuse Center (WDRC), King Abdullah University of Science and Technology.
peng.wang@kaust.edu.sa

Renyuan Li
Biological and Environmental Sciences and Engineering Division, Water Desalination and Reuse Center (WDRC), King Abdullah University of Science and Technology.
peng.wang@kaust.edu.sa

Swasmi Purwajanti
Australian Institute for Bioengineering and Nanotechnology (AIBN), Corner College and Cooper Rds (Bldg 75), The University of Queensland, Brisbane Qld 4072, Australia.
c.yu@uq.edu.au

Volodymyr V. Tarabara
Department of Civil and Environmental Engineering, Michigan State University, East Lansing 48824 USA.
tarabara@egr.msu.edu

Xiaodan Huang
Australian Institute for Bioengineering and Nanotechnology (AIBN), Corner College and Cooper Rds (Bldg 75), The University of Queensland, Brisbane Qld 4072, Australia.
c.yu@uq.edu.au

Yuxiong Huang
Bren School of Environmental Science and Management, University of California at Santa Barbara, CA, USA 93106.
keller@bren.ucsb.edu

Preface

Water pollution and water scarcity are among the most challenging problems facing mankind nowadays. With rapid population growth, steadily improving life standards, fast industrialization and modernization of developing countries, these challenges will persist, if not worsen, in the years to come. With conventional water treatment technologies being pushed towards their capacity limits, it is now a popular perception that the solutions to the existing and future water challenges will hinge upon further developments in nanotechnology.

Ever since 1959, when the term "nanotechnology" was first used by Richard Feynman in his famous lecture entitled "there's plenty of room at the bottom", the field of nanotechnology has been experiencing literally explosive growth, especially in the last two decades. Moreover, the application of nanotechnology to water treatment has steadily grown into a distinct field with the expected growth rate on an exponential rise.

In the early days when nanomaterials first attracted attention from researchers in water field, trial-and-error approach prevailed in which water scientists searched suitable applications for the nanomaterials developed by materials scientists and the disconnection between two sides was common. The rational design concept came into being when the researchers realized that the chemistry and ultimately the functions of nanomaterials could be deliberately pre-designed for a desired purpose before embarking on nanomaterial synthesis. Within the scheme of rational design, material design, synthesis and application are seamlessly integrated within one entity. The rational nano-design starts with a clear problem definition, necessitates interdisciplinary approaches, involves 'think-outside-the-box', and represents the future growth point of water field. However, it is still largely new to the educated public and even scientists and engineers in water fields. Therefore, it is the purpose of this book to promote the concept of rational nano-design and to demonstrate its creativity, innovation, and excitement in water treatment. The book presents a series of carefully selected rationally designed nano-materials/devices/surfaces to embody the concept of nano-design and to illustrate its remarkable potential to change the face of the research in water industry in the future. The selected examples in the book chapters represent drastically different, ground-breaking, and eye-opening approaches to conventional problems and each of the book contributors is world-renowned expert in the burgeoning field of rational nano-design for applications.

As you will see, the topics of the book chapters are truly multidisciplinary. They span from an introduction to rational nano-design for water applications (Chapter 1, Renyuan Li and Professor Peng Wang from KAUST), design and application of magnetic-core composite nano/micro particles for environmental remediation (Chapter 2, Yuxiong Huang and Professor Arturo Keller from University of California, Santa Barbara), rational design of functional nanoporous materials to confine water pollutant in controlled nano-space (Chapter 3, Swasmi Purwajanti, Jie Yang, Xiaodan Huang, and Professor Chengzhong Yu from University of Queensland), hierarchical materials as a design concept for multifunctional membranes, (Chapter 4, Christopher Crock, Brian Starr, and Professor Volodymyr Tarabara from Michigan State University), smart membrane materials for controllable oil-water separation (Chapter 5, Lianbin Zhang and Professor Peng Wang from KAUST), design of the next-generation FO draw solution (Chapter 6, Dr. Aaron Wilson from Idaho National Laboratory, USA), to nanotechnology for microbial fuel cells (Chapter 7, Professor Muhammad Mustafa Hussain from KAUST).

Rational Design of Next-generation Nanomaterials and Nanodevices for Water Applications is intended for undergraduates, graduates, scientists and professionals in the fields of environmental science, material science, chemistry, and chemistry engineering. It provides coherent and good material for teaching, research, and professional reference. I hope that this book will provide an inspiration for readers who are interested in rational design of nanotechnology and who are passionate at further exploring nanomaterials to make contributions to the solutions to our grand environmental challenges.

Peng Wang
Water Desalination and Reuse Center
King Abdullah University of Science and Technology
Thuwal, Saudi Arabia

Chapter 1

Introduction to rational nano-design for water applications

Renyuan Li and Peng Wang

In the past century, the development in water treatment technologies has made critical contribution to sustaining human society. Bulk water chemistry-based conventional methods, such as adsorption (Ali & Gupta, 2007; Le Cloirec *et al.* 1997), advanced oxidization (Liu *et al.* 2007; Vilve *et al.* 2009), bio-treatment (Lazarova & Manem, 1995; Lettinga *et al.* 1980), have been widely utilized at industrial scale for providing quality water for societal benefits. At the same time, with the ever-growing human population and also ever-increasing life quality expectation by each individual, the human society has been pushing the water and energy demand to the limit of environment capacity (Hanjra & Qureshi, 2010; Barnham *et al.* 2006).

The world energy consumption is 15TW nowadays and it is expected to increase to 30TW by 2050 (Zhang & Wang, 2012). Sadly, the global energy consumption is and will be, for the foreseeable future, heavily dominated by burning non-renewable fossil fuels, especially coal and oil, whose stock is shrinking at an alarming pace and whose usage leads to negative environment impact, majorly environmental pollution and global warming (McCollum *et al.* 2013) (Figure 1.1). It is true that the water pollution along with water scarcity are becoming severer in many parts of the world as a result of increasing energy consumption by these regions. It is now a popular notion that water security and energy security are two intricately intertwined two grand challenges of our times (Jacobson, 2009; Vorosmarty *et al.* 2010; Grey & Sadoff, 2007), with neither of which can be solved without looking at the other.

The design of energy and water systems based on bulk chemistry experienced its golden age in the first half of 20 century and has gradually reached their steady states, which propelled research attentions to smaller scales then to beat science hard to show its best to meet the human demand. The concept of 'nano' came into being naturally then. In 1959, Richard Feynman first used the term "nanotechnology" in his famous lecture entitled "there's plenty of room at the bottom", which is hailed by many as the herald of the era of nano (Feynman, 1992).

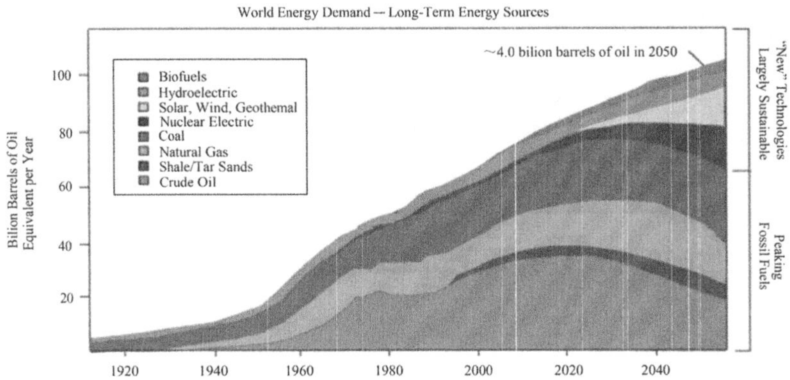

Figure 1.1 Scheme of long term energy sources for world energy demand. Source: Lynn Orr, *Changing the world energy systems*, Global Climate & Energy Project (GCEP).

Nanomaterials have two primary advantages over conventional bulk materials: (1) they have small size and thus big specific surface area, which are beneficial to many interface-related applications; (2) their properties, including chemical, physical, optical, electronic, mechanical, and magnetic properties, can be judiciously tuned by controlling their size, surface morphology, shape and crystal orientation, etc. As a result, going to nanoscale has opened up numerous new avenues that would otherwise be impossible with conventional bulk materials. With a loose definition of nanomaterial being the ones with controllable features at nanometer scale, the general field of nanomaterials has been experiencing literally explosive growth especially during the past two decades and the field of nanomaterial for water applications is no exception (Figure 1.2).

Approach wise, the field of nanomaterial for water applications experienced two two distinct development stages: (1) trial-and-error stage where efforts were made in searching suitable applications in water treatment for the nanomaterials developed by material scientists in a trial-and-error manner. (2) Rational design stage at which nanomaterial design is initiated only based on a scientifically clearly defined problem definition. Within the first stage, water chemists and material scientists tended to work separately and alone and interdisciplinary cross-conversation was rare. By the end of the trial-and-error stage, researchers came to realize that chemistry and ultimately the functions of their nanomaterials could be deliberately pre-designed for a desired purpose. At the rational design stage, the focus is on 'design-for-purpose' before embarking nanomaterial synthesis (Figure 1.3).

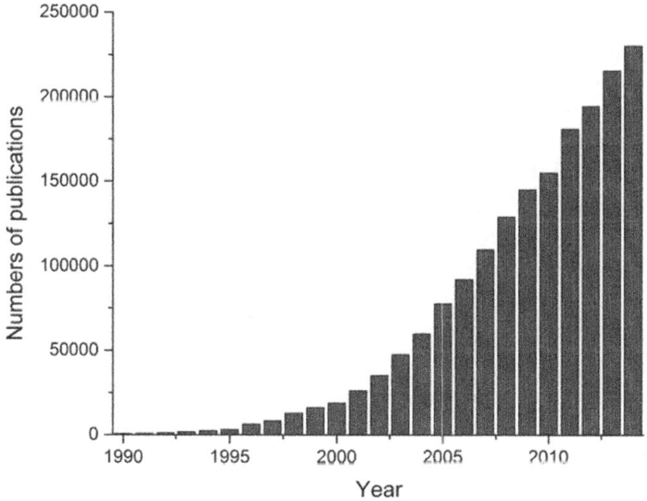

Figure 1.2 Steady growth of annual publication number in nano field in the last 3 decades. The annual publication # was 704 in 1990 while 230174 in 2014. Source from Web of Science, by searching the topic key words *nano** on January 16th, 2016.

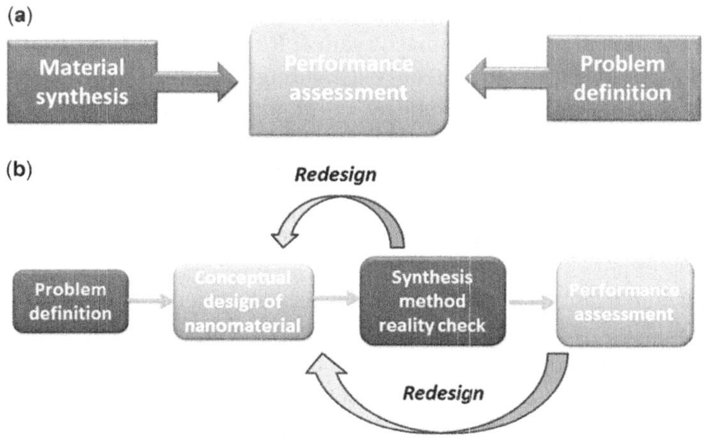

Figure 1.3 Schematic illustration of (a) Trial-and-error approach and (b) rational design of nanomaterial. Reprinted from reference (Li *et al.* 2015) – Copyright The Royal Society of Chemistry.

In more details, unlike the trial-and-error approach, a rational design process starts with scientifically, generally chemically, defining the problem to be solved. Based on the clear problem definition, a conceptual design of a nanomaterial-based solution is proposed, fed back to the problem definition for scientific check. The communication is iterated until both the problem definition and nanomaterial design agree well with each other. Next, the conceptually designed nanomaterial is checked against the currently available synthesis capability and can then be synthesized if available. Otherwise, the iteration back to the nanomaterial design will take place until the designed nanomaterial can be successfully synthesized. The performance of the synthesized nanomaterial is then assessed with respect to its design purpose, which has been unambiguously defined in the problem definition step and the iteration back to the nanomaterial design will take place again in the event of an unsatisfactory performance of the nanomaterial.

Due to the space limit, the chapter restrains itself from providing more detailed discussion on the concept of 'rational design of nanomaterial for water treatment' and interested readers are encouraged to consult our recent review article on the topic (Li *et al.* 2015).

There have been numerous exciting developments in the field, but only selected examples are highlighted in this chapter.

1.1 RATIONAL DESIGN OF MAGNETIC NANOMATERIALS AS ADSORBENTS FOR WATER TREATMENT

Adsorption has long been developed as one of well-established water treatment methods (Lambert *et al.* 1996; Namasivayam & Ranganathan, 1995; Namasivayam & Kavitha, 2002). Designing of an outstanding adsorbent should consider its adsorption capacity, selectivity, stability, reusability, recoverability, and economic feasibility. The development of nanoporous, especially well-ordered mesoporous, materials represents a significant milestone in adsorption as these nanoporous adsorbents possess very high surface areas, large and regularly ordered mesoscale channels, and fast mass transfer kinetics (Wan *et al.* 2008; Zhang *et al.* 2009; Yang *et al.* 2014).

Another important factor influencing adsorbent performance is the interaction between the active sites on the adsorbents and the targeted adsorbates, which is obtained by surface chemical modification of the adsorbents (Zhao & Lu, 1998). The surface chemistry controls the selectivity and strength of the adsorption. Fortunately, many of the nanoporous adsorbents are compatible with a wide range of chemical modifications. As a result, surface chemical functionalization especially of nanoporous materials is a basis of many rational designs of effective adsorbents for water treatment (Walcarius & Mercier, 2010; Feng *et al.* 1997).

In practical applications, separation and subsequent recycling of the adsorbents is essential from the operation cost point of view. Filtration, centrifugation, sedimentation are among the conventional separation approaches (Wang *et al.* 2009). Recently, magnetic field induced separation of magnetic adsorbents are emerging as a cost-effective separation method in water treatment (Wang *et al.* 2010a; Zhang & Kong, 2011).

The advantage of utilizing magnetic separation is obvious particularly in water treatment processes because: (1) it has a lower energy consumption than other conventional methods such as centrifugation especially at large scales; (2) it overcomes fouling and clogging typically occurring in filtration processes; (3) kinetic of magnetic separation is controllable by the intensity of external magnetic field and can be much faster than gravity based sedimentation.

In principle, magnetic properties can be endowed to many conventional adsorbents. One example is Magnetic Ion Exchange (MIEX) resin, in which ion-exchange resin is loaded with high content of magnetic iron oxide particles. Despite the small size (around 180 μm in diameter), the resin beads provide a large number of adsorption sites and these magnetized resin beads work as weak individual magnets and tend to form agglomerates, which further lead to a rapid settling or fluidize with a high hydraulic loading rates (Singer & Bilyk, 2002). Currently MIEX resins for both anion or cation exchanges are available. Figure 1.4 illustrates an example using MIEX resin for the removal of dissolved organic carbon (DOC) where the adsorption site on MIEX resin is chloride. In regeneration process, the DOC loaded MIEX resin is placed in a concentrated NaCl solution to achieve a reverse ion exchange in which the resin releases DOC and the chloride retakes adsorption site in the resin (MIEX).

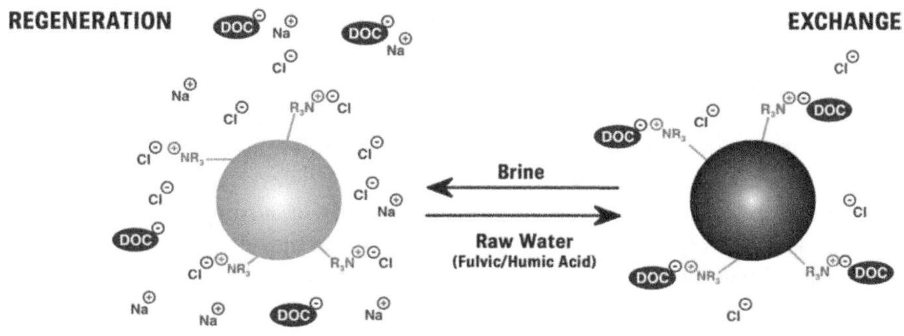

Figure 1.4 Schematic of removal of DOC by MIEX and its regeneration process. Source: MIEX website, IXOM Application bulletin.

Generally, magnetic adsorbents can have two different configurations: (1) magnetic component serving as adsorbent (Ai *et al.* 2008; Hu *et al.* 2005); (2) magnetic component providing nothing more than magnetic separation mechanism with the adsorbing component being something else (Yao *et al.* 2012). The example of the former can be mesoporous γ-Fe_2O_3 nanoparticles for chromium (VI) removal reported by Wang *et al.*, in which γ-Fe_2O_3 is both adsorbing component and magnetic component (Wang & Lo, 2009). One example of the latter can be core-shell structured magnetic permanently confined micelle arrays reported by the same group for hydrophobic organic contaminant removal (Wang *et al.* 2009). The magnetic core of the material does not contribute to the adsorption of the contaminant, but to provide separation means by responding to magnetic field. More detailed discussion on magnetic adsorbents can be found in Chapters 2 and 3.

1.2 RATIONAL DESIGN OF SUPERWETTING MEMBRANE FOR OIL-WATER SEPARATION

The effective and quick removal of accidentally spilled oil in the environment is essential to minimize its adverse environmental impact. Traditional oil spilling responses, including physical skimmers, hydrocyclone based separation, adsorption, face the challenge of low recovery efficiency, high-energy consumption, and high cost (Wang *et al.* 2015; Zhu *et al.* 2014). Materials especially membranes based on differentiating surface wetting behaviors between oil, water, have recently attracted considerably research interests (Chu *et al.* 2015).

For such a material/membrane to be successful, there are many aspects that should be considered into the material design: 1) physic and chemical properties of targeted oil that need be separated from water. 2) Separation material properties, which is essential for the minimization of the challenges such as clogging, fouling and chemical etching. 3) Elongation of membrane or other oil/water separation material lifetime, such as endowing them with self-cleaning or self-healing properties. As an example, Figure 1.5 shows the concept of a self-cleaning underwater superoleophobic mesh for oil-water separation. Underwater superoleophobic material usually possesses a superhydrophillic surface. When a superhydrophillic material is immersed into water, a water layer forms on the material surface, which prevents its contact with oil. However, once the underwater superoleophobic material is contaminated by oil via adsorption of dissolved species, the material losses its oleophobic property and thus its ability to separation oil/water. To overcome this problem, TiO_2 was inducted into the surface coating layer on stainless steel mesh through lay-by-layer assembly method and it endowed the separation mesh a self-cleaning function under the illumination of UV light. The TiO_2 helped to decompose the adsorbed fouling species and thus recovered the oleophobicity of the mesh (Zhang *et al.* 2013) (Figure 1.5).

The Chapter 5 of this book is devoted to rational designing of nanomaterial for controllable oil-separation processes.

1.3 EMERGING NANO-BASED NEXT GENERATION MEMBRANES

The development of conventional membrane based separation, including microfiltration (MF), nanofiltration (NF), ultrafiltration (UF), forward osmosis

(FO), reverse osmosis (RO) and membrane distillation (MD), and new nano-assisted membrane processes have recorded a significant progress during past decade (Lee *et al.* 2011; Hegab & Zou, 2015; Pendergast & Hoek, 2011; Crock *et al.* 2013). The discovery and development of nanomaterials, often aided by MD simulation nanotechnology open a lot of new avenue beyond the limits of conventional membrane materials (Kalra *et al.* 2003; Cohen-Tanugi & Grossman, 2012; Walther *et al.* 2013). The recent rise of two-dimensional nanomaterials especially carbon-based nanomaterials, such as graphene, graphene derivatives, carbon nanotubes, provide the membrane fields with a lot of new possibilities and present a very promising direction in next-generation water treatment.

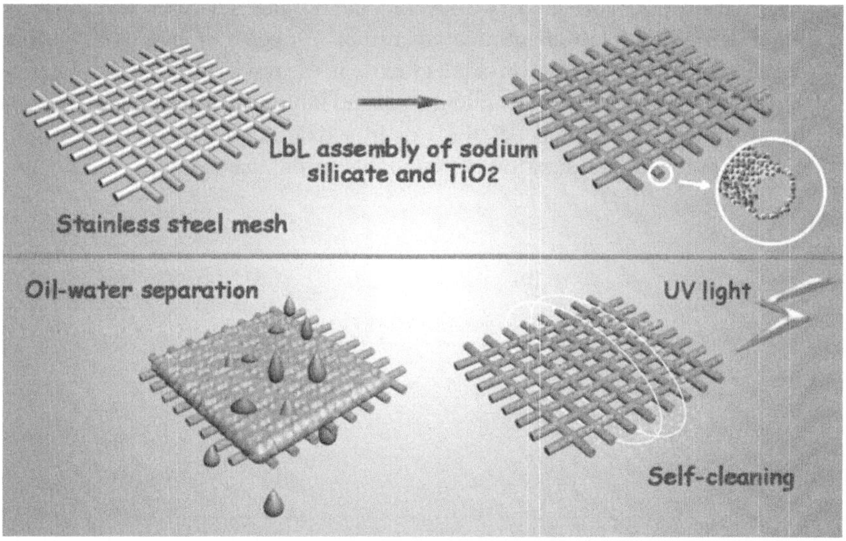

Figure 1.5 TiO_2 coated underwater superoleophobic mesh for oil-water separation and its self-cleaning process Source: Wang *et al.* A self-cleaning underwater superoleophobic mesh for oil-water separation.

Alongside with membrane material development, researchers also devote their passion in making synergistically multi-functional all-in-one membranes, many of which represent proof-of-concepts of some ground-breaking and next-generation concepts in water treatment. Comparing with the conventional concept of membranes being permeable and physical barriers for the physical separation of two bulk phases, active filtration is emerging as a more cost-effective alternative wherein adsorption and/or more importantly chemical processes (*e.g.* catalysis, reduction and oxidization) are being combined with membrane filtration to achieve active filtration for a more energy-efficient water treatment. When operated under a pressure-driven convective flow, the active membranes provide reactants with rapid access to active sites, thereby minimizing the mass transfer limitations associated with other high surface area-to-volume materials and leading to enhanced treatment performance. Examples of active filtration membranes include TiO_2 or other semiconductor-based photo (electrochemical)-catalytically active membrane filters (Albu *et al.* 2007), noble-metal (e.g. Au and Pd) based catalytically active membrane filters (Wang *et al.* 2013), CNT-based electrochemically active filters (Vecitis *et al.* 2011) and Fenton-reaction assisted reactive membrane filters (Julbe *et al.* 2001).

In 2007, Bhattacharyya *et al.* reported a rationally designed all-in-one Fenton-reaction-active filtration system (Figure 1.6), which integrated within one platform nanostructured materials, enzymatic catalysis and iron-catalyzed free radical reactions. Within the fabricated two-layered membrane, glucose oxidase (GOx) was immobilized in the top membrane layer to *in situ* generate hydrogen peroxide through reacting with deliberately added glucose in the raw water. The as-generated hydrogen peroxide was flushed down to the second layer membrane where it reacted with the polymer-immobilized Fe^{2+}/Fe^{3+} or iron oxide nanoparticles to initialize Fenton reaction for the oxidization of pollutants in the water within the confined space (Lewis *et al.* 2011).

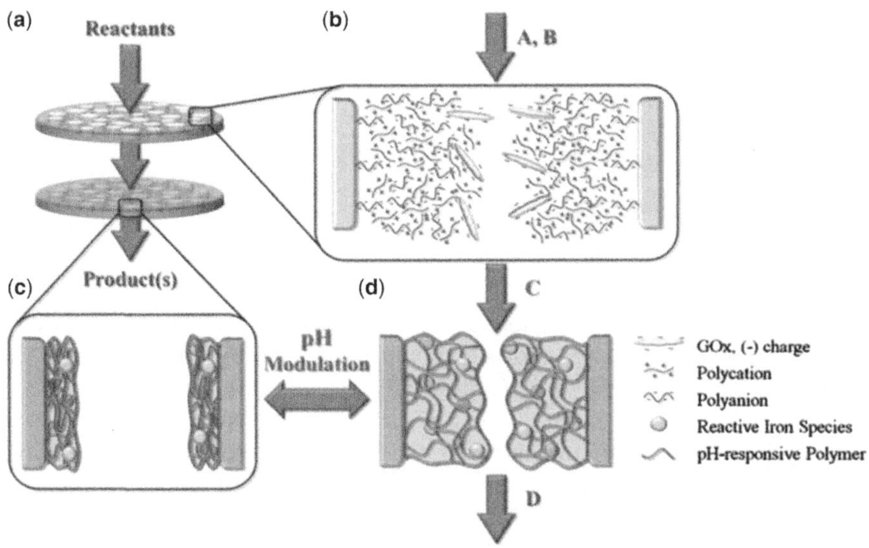

Figure 1.6 (a) Schematic of two-layered all-in-one Fenton-reaction-active filtration system. (b) Pore space of the top layer membrane contains bio-active enzyme GOx to generate H_2O_2 using glucose. (c) and (d) Pore space of the second layer which is pH responsive and contains reactive iron species to generate free radical using the H_2O_2 coming from first layer. Source: Bhattacharyya *et al.* Reactive nanostructured membranes for water purification.

The Chapter 4 covers the multifunctional active filtration membranes using hierarchical materials as a design concept.

1.4 RATIONAL DESIGN OF FO DRAW SOLUTION

Unlike reverse osmosis (RO) process, forward osmosis (FO) is a naturally occurring process where water in lower salinity side permeates through a semi-permeable membrane and moves to the side with higher salinity due to the osmosis pressure difference (Yip *et al.* 2010; Wang *et al.* 2010b). FO membrane is similar to RO membrane in configuration, with both containing a thin dense layer for salt-water separation and a thick support layer (Lee *et al.* 2011; Lau *et al.* 2012). Comparing with RO process which requires a high pressure to overcome natural osmosis pressure to force water molecules through membrane, FO process, due to its require needs no external energy input and has find many niche applications in both water desalination and wastewater treatment (Cath *et al.* 2006; Zhao *et al.* 2012) (Figure 1.7).

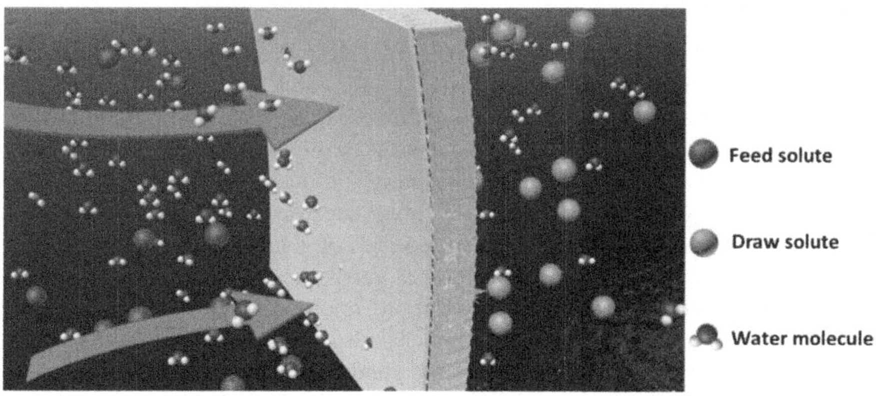

Figure 1.7 Schematic of FO process. Water molecules penetrate through FO dense layer from feed solution side to draw solution side spontaneously due to the osmosis pressure difference.

In parallel to the research efforts in improving FO membrane material's performance, considerable efforts have also been invested in designing novel draw solution to enhance the FO system energy efficiency (McCutcheon *et al.* 2006; Achilli *et al.* 2010; Ge *et al.* 2013). Ideally, a well-designed FO draw solution should generate a high osmotic pressure and lower down the reverse flux of draw solute to enhance the permeate flux and ensure rejection rate. Meanwhile, the draw solution should be easily regenerated after use, and be non-toxic and non-corrosive. One of the commonly used draw solution for FO is inorganic salts, which was first developed in 1970s by Hough *et al.* (Hough, 1970). The use of salt as draw solution requires further treatment to recover permeate and to regenerate draw solution, which entails additional energy consumption.

In the past decade, attempts have been made in replacing conventional inorganic salt with some nanomaterials in the FO draw solution with certain success. The concept of utilizing hydrophilic magnetic nanoparticles as draw solutes was firstly proposed by Warne *et al.* in 2008 (Warne *et al.* 2008). Hydrophilicity endowed highly dispersive property of nanoparticle while magnetic property made them easy to be separated through external magnetic field after use. However, due to large particle size and suspension state of these nanoparticles in water, the osmotic pressure generated by the nanoparticles is generally not as high as salt solution, which can be translated to lower permeate flux and lower membrane efficiency. The Chapter 6 of the book will discuss the recent developments in FO draw solution design.

1.5 RATIONAL DESIGNED MICRO-SIZED MICROBIAL FUEL CELL FOR HIGHLY EFFICIENCY ENERGY HARVESTING

The human knowledge that bacteria can generate electric current has a long history and can be traced back to 1911 (Pant *et al.* 2010). Research interest on microbial fuel cell (MFC), where the energy contained within organic waste such as biomass and domestic wastewaters is converted into electricity, has being growing steadily in the past decade. It is believed by some that the energy produced by microbes can be the most important energy source in the near future because MFC provides a possible way to recover electric energy from a wide series of soluble or dissolved complex organic wastes (Santoro *et al.* 2016; Rabaey & Verstraete, 2005; Logan *et al.* 2006).

In a MFC, under an anaerobic condition, microorganisms oxidize organic matters as their food in the anode chamber and generate electrons and protons. The produced electrons immigrate from the anode to cathode through an external circuit, combine with electron acceptor (e.g., oxygen) on the cathode to produce water. There are two types of system configurations of MFC: namely single-chamber (Reddy *et al.* 2010; Venkata Mohan *et al.* 2008) and two-chamber (Xiao *et al.* 2011; Nien *et al.* 2011) MFCs. The major difference between the two is the presence of a semi-permeable and ion-selective membrane that separates the cathode and anode chambers in the two-chamber configuration.

Besides the structural configuration, other aspects such as mass transportation, reaction kinetics and fabrication cost all influence the performance and applications of MFCs. The mass transfer and reaction kinetics are affected by several factors in the systems, including: 1) flow rates, 2) solutes in both chamber, 3) morphology of electrodes (i.e. size and shape), 4) available area for solutes transportation between anode and cathode, 5) electrode materials, 6) biofilm conditions and 7) surface properties of electrodes (Wang *et al.* 2011). Given the factors described above, micro-sized microbial fuel cell, especially µl-scale MFCs, due to their large surface area-to-volume ratio, low Reynolds number, fast response time and short electrode distance, provide a highly efficient way for on-chip energy harvesting.

One example of micro-sized MFC is shown in Figure 1.8. This micro-sized MFC contains µl sized chamber (i.e. 1.5 µl for anode chamber and 4 µl for cathode chamber), use Fe(III) solution as electron acceptor, Shewanella oneidensis strain MR-1 as electronic generation microorganism (Qian *et al.* 2009).

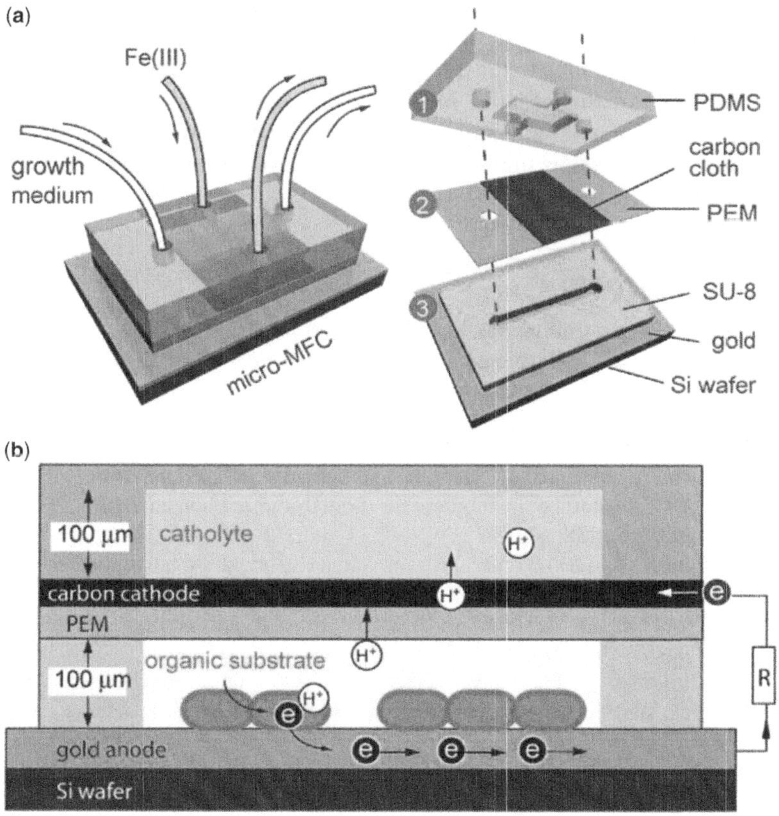

Figure 1.8 (a) Schematic of Micro-sized MFC. Left: The arrow indicates the microfluidic flow pathway directions, Fe(III) served as electron acceptor media and growth medium to support the life cycle of bacteria. Right: The structure of MFC. The cell was build and fabricated in a silicon wafer, PDMS and SU-8 photo resist served as cell chamber. Carbon cloth on the proton exchange membrane served as cathode, gold film served as anode. (b) Illustration of micro-sized MFC working principle. The microbial was immobilized on gold anode utilize growth medium generate protons and electrons. The protons transfer to catholyte through proton exchange membrane while electron transfer to carbon cathode through external circuit. Source: Qian *et al.* A 1.5 µL microbial fuel cell for on-chip bioelectricity generation.

The Chapter 7 of the book will discussed *'Rational designed micro-sized microbial fuel cell for highly efficiency energy harvesting'*.

1.6 CONCLUSION

Given the nature of rational design, rational design involves "thinking-outside-the-box", is not bounded by the available nanomaterials, and thus has a high potential of creating next-generation and ground-breaking solutions to the water challenges of our times. In theory, any material developed within the scheme of rational design is new and thus contributes to the already vast library of nanomaterials.

In view of the amazing progress in the field during past decade, we are confident that the rational design of nanomaterial will carry on evolving and provide us even more unprecedented opportunities to solve the environment and energy crisis throughout a sustainable ways.

1.7 REFERENCES

Achilli A., Cath T. Y. and Childress A. E. (2010). Selection of inorganic-based draw solutions for forward osmosis applications. *Journal of Membrane Science*, **364**(1–2), 233–241.

Ai Z., Cheng Y., Zhang L. and Qiu J. (2008). Efficient removal of Cr(VI) from aqueous solution with Fe@Fe2O3 core–shell nanowires. *Environmental Science & Technology*, **42**(18), 6955–6960.

Albu S. P., Ghicov A., Macak J. M., Hahn R. and Schmuki P. (2007). Self-organized, free-standing TiO2 nanotube membrane for flow-through photocatalytic applications. *Nano Letters*, **7**(5), 1286–1289.

Ali I. and Gupta V. K. (2007). Advances in water treatment by adsorption technology. *Nature Protocols*, **1**(6), 2661–2667.

Barnham K. W. J., Mazzer M. and Clive B. (2006). Resolving the energy crisis: nuclear or photovoltaics?. *Nature Materials*, **5**(3), 161–164.

Cath T. Y., Childress A. E. and Elimelech M. (2006). Forward osmosis: principles, applications, and recent developments. *Journal of Membrane Science*, **281**(1–2), 70–87.

Chu Z., Feng Y. and Seeger S. (2015). Oil/water separation with selective superantiwetting/superwetting surface materials. *Angewandte Chemie International Edition*, **54**(8), 2328–2338.

Cohen-Tanugi D. and Grossman J. C. (2012). Water desalination across nanoporous graphene. *Nano Letters*, **12**(7), 3602–3608.

Crock C. A., Rogensues A. R., Shan W. and Tarabara V. V. (2013). Polymer nanocomposites with graphene-based hierarchical fillers as materials for multifunctional water treatment membranes. *Water Research*, **47**(12), 3984–3996.

Feng X., Fryxell G. E., Wang L.-Q., Kim A. Y., Liu J. and Kemner K. M. (1997). Functionalized monolayers on ordered mesoporous supports. *Science*, **276**(5314), 923–926.

Feynman R. P. (1992). There's plenty of room at the bottom [data storage]. *Journal of Microelectromechanical Systems*, **1**(1), 60–66.

Ge Q., Ling M. and Chung T.-S. (2013). Draw solutions for forward osmosis processes: developments, challenges, and prospects for the future. *Journal of Membrane Science*, **442**, 225–237.

Grey D. and Sadoff C. W. (2007). Sink or swim? Water security for growth and development. *Water Policy*, **9**(6), 545–571.

Hanjra M. A. and Qureshi M. E. (2010). Global water crisis and future food security in an era of climate change. *Food Policy*, **35**(5), 365–377.

Hegab H. M. and Zou L. (2015). Graphene oxide-assisted membranes: fabrication and potential applications in desalination and water purification. *Journal of Membrane Science*, **484**, 95–106.

Hough W. T. (1970). Process for Extracting Solvent from a Solution. US Patent 3532621.

Hu J., Chen G. and Lo I. M. C. (2005). Removal and recovery of Cr(VI) from wastewater by maghemite nanoparticles. *Water Research*, **39**(18), 4528–4536.

Jacobson M. Z. (2009). Review of solutions to global warming, air pollution, and energy security. *Energy & Environmental Science*, **2**(2), 148–173.

Julbe A., Farrusseng D. and Guizard C. (2001). Porous ceramic membranes for catalytic reactors — overview and new ideas. *Journal of Membrane Science*, **181**(1), 3–20.

Kalra A., Garde S. and Hummer G. (2003). Osmotic water transport through carbon nanotube membranes. *Proceedings of the National Academy of Sciences*, **100**(18), 10175–10180.

Lambert T. W., Holmes C. F. B. and Hrudey S. E. (1996). Adsorption of microcystin-LR by activated carbon and removal in full scale water treatment. *Water Research*, **30**(6), 1411–1422.

Lau W. J., Ismail A. F., Misdan N. and Kassim M. A. (2012). A recent progress in thin film composite membrane: a review. *Desalination*, **287**, 190–199.

Lazarova V. and Manem J. (1995). Biofilm characterization and activity analysis in water and wastewater treatment. *Water Research*, **29**(10), 2227–2245.

Le Cloirec P., Brasquet C. and Subrenat E. (1997). Adsorption onto Fibrous Activated Carbon: Applications to Water Treatment. *Energy & Fuels*, **11**(2), 331–336.

Lee K. P., Arnot T. C. and Mattia D. (2011). A review of reverse osmosis membrane materials for desalination—Development to date and future potential. *Journal of Membrane Science*, **370**(1–2), 1–22.

Lettinga G., van Velsen A. F. M., Hobma S. W., de Zeeuw W. and Klapwijk A. (1980). Use of the upflow sludge blanket (USB) reactor concept for biological wastewater treatment, especially for anaerobic treatment. *Biotechnology and Bioengineering*, **22**(4), 699–734.

Lewis S. R., Datta S., Gui M., Coker E. L., Huggins F. E., Daunert S., Bachas L. and Bhattacharyya D. (2011). Reactive nanostructured membranes for water purification. *Proceedings of the National Academy of Sciences*, **108**(21), 8577–8582.

Li R., Zhang L. and Wang P. (2015). Rational design of nanomaterials for water treatment. *Nanoscale*, **7**(41), 17167–17194.

Liu H., Wang C., Xiangzhong, Xuan X., Jiang C. and Cui H. N. (2007). A novel electro-Fenton process for water treatment: reaction-controlled pH adjustment and performance assessment. *Environmental Science & Technology*, **41**(8), 2937–2942.

Logan B. E., Hamelers B., Rozendal R., Schröder U., Keller J., Freguia S., Aelterman P., Verstraete W. and Rabaey K. (2006). Microbial fuel cells: methodology and technology. *Environmental Science & Technology*, **40**(17), 5181–5192.

McCollum D., Bauer N., Calvin K., Kitous A. and Riahi K. (2013). Fossil resource and energy security dynamics in conventional and carbon-constrained worlds. *Climatic Change*, **123**(3), 413–426.

McCutcheon J. R., McGinnis R. L. and Elimelech M. (2006). Desalination by ammonia–carbon dioxide forward osmosis: influence of draw and feed solution concentrations on process performance. *Journal of Membrane Science*, **278**(1–2), 114–123.

MIEX Dissolved Organic Carbon Removal for DBP Reduction & Improved Process Performance. IXOM Application bulletin.

Namasivayam C. and Kavitha D. (2002). Removal of Congo Red from water by adsorption onto activated carbon prepared from coir pith, an agricultural solid waste. *Dyes and Pigments*, **54**(1), 47–58.

Namasivayam C. and Ranganathan K. (1995). Removal of Cd(II) from wastewater by adsorption on "waste" Fe(III)Cr(III) hydroxide. *Water Research*, **29**(7), 1737–1744.

Nien P.-C., Lee C.-Y., Ho K.-C., Adav S. S., Liu L., Wang A., Ren N. and Lee D.-J. (2011). Power overshoot in two-chambered microbial fuel cell (MFC). *Bioresource Technology*, **102**(7), 4742–4746.

Pant D., Van Bogaert G., Diels L. and Vanbroekhoven K. (2010). A review of the substrates used in microbial fuel cells (MFCs) for sustainable energy production. *Bioresource Technology*, **101**(6), 1533–1543.

Pendergast M. M. and Hoek E. M. V. (2011). A review of water treatment membrane nanotechnologies. *Energy & Environmental Science*, **4**(6), 1946–1971.

Qian F., Baum M., Gu Q. and Morse D. E. (2009). A 1.5 μL microbial fuel cell for on-chip bioelectricity generation. *Lab on a Chip*, **9**(21), 3076–3081.

Rabaey K. and Verstraete W. (2005). Microbial fuel cells: novel biotechnology for energy generation. *Trends in Biotechnology*, **23**(6), 291–298.

Reddy M. V., Srikanth S., Mohan S. V. and Sarma P. N. (2010). Phosphatase and dehydrogenase activities in anodic chamber of single chamber microbial fuel cell (MFC) at variable substrate loading conditions. *Bioelectrochemistry*, **77**(2), 125–132.

Santoro C., Soavi F., Serov A., Arbizzani C. and Atanassov P. (2016). Self-powered supercapacitive microbial fuel cell: the ultimate way of boosting and harvesting power. *Biosensors and Bioelectronics*, **78**, 229–235.

Singer P. C. and Bilyk K. (2002). Enhanced coagulation using a magnetic ion exchange resin. *Water Research*, **36**(16), 4009–4022.

Vecitis C. D., Schnoor M. H., Rahaman M. S., Schiffman J. D. and Elimelech M. (2011). Electrochemical multiwalled carbon nanotube filter for viral and bacterial removal and inactivation. *Environmental Science & Technology*, **45**(8), 3672–3679.

Venkata Mohan S., Veer Raghavulu S. and Sarma P. N. (2008). Biochemical evaluation of bioelectricity production process from anaerobic wastewater treatment in a single chambered microbial fuel cell (MFC) employing glass wool membrane. *Biosensors and Bioelectronics*, **23**(9), 1326–1332.

Vilve M., Hirvonen A. and Sillanpää M. (2009). Effects of reaction conditions on nuclear laundry water treatment in Fenton process. *Journal of Hazardous Materials*, **164**(2–3), 1468–1473.

Vorosmarty C. J., McIntyre P. B., Gessner M. O., Dudgeon D., Prusevich A., Green P., Glidden S., Bunn S. E., Sullivan C. A., Liermann C. R. and Davies P. M. (2010). Global threats to human water security and river biodiversity. *Nature*, **467**(7315), 555–561.

Walcarius A. and Mercier L. (2010). Mesoporous organosilica adsorbents: nanoengineered materials for removal of organic and inorganic pollutants. *Journal of Materials Chemistry*, **20**(22), 4478–4511.

Walther J. H., Ritos K., Cruz-Chu E. R., Megaridis C. M. and Koumoutsakos P. (2013). Barriers to superfast water transport in carbon nanotube membranes. *Nano Letters*, **13**(5), 1910–1914.

Wan Y., Shi Y. and Zhao D. (2008). Supramolecular aggregates as templates: ordered mesoporous polymers and carbons. *Chemistry of Materials*, **20**(3), 932–945.

Wang B., Liang W., Guo Z. and Liu W. (2015). Biomimetic super-lyophobic and super-lyophilic materials applied for oil/water separation: a new strategy beyond nature. *Chemical Society Reviews*, **44**(1), 336–361.

Wang H., Dong Z. and Na C. (2013). Hierarchical carbon nanotube membrane-supported gold nanoparticles for rapid catalytic reduction of p-nitrophenol. *ACS Sustainable Chemistry & Engineering*, **1**(7), 746–752.

Wang H.-Y., Bernarda A., Huang C.-Y., Lee D.-J. and Chang J.-S. (2011). Micro-sized microbial fuel cell: a mini-review. *Bioresource Technology*, **102**(1), 235–243.

Wang J., Zheng S., Shao Y., Liu J., Xu Z. and Zhu D. (2010a). Amino-functionalized Fe3O4@SiO2 core–shell magnetic nanomaterial as a novel adsorbent for aqueous heavy metals removal. *Journal of Colloid and Interface Science*, **349**(1), 293–299.

Wang P. and Lo I. M. C. (2009). Synthesis of mesoporous magnetic γ-Fe2O3 and its application to Cr(VI) removal from contaminated water. *Water Research*, **43**(15), 3727–3734.

Wang P., Shi Q., Shi Y., Clark K. K., Stucky G. D. and Keller A. A. (2009). Magnetic permanently confined micelle arrays for treating hydrophobic organic compound contamination. *Journal of the American Chemical Society*, **131**(1), 182–188.

Wang R., Shi L., Tang C. Y., Chou S., Qiu C. and Fane A. G. (2010b). Characterization of novel forward osmosis hollow fiber membranes. *Journal of Membrane Science*, **355**(1–2), 158–167.

Warne B., Buscall R., Mayes E., Oriard T. and Norris I. (2008). Water purification method. GB patent 2464956.

Xiao B., Yang F. and Liu J. (2011). Enhancing simultaneous electricity production and reduction of sewage sludge in two-chamber MFC by aerobic sludge digestion and sludge pretreatments. *Journal of Hazardous Materials*, **189**(1–2), 444–449.

Yang J., Zhang H., Yu M., Emmanuelawati I., Zou J., Yuan Z. and Yu C. (2014). High-content, well-dispersed γ-Fe_2O_3 nanoparticles encapsulated in macroporous silica with superior arsenic removal performance. *Advanced Functional Materials*, **24**(10), 1354–1363.

Yao Y., Miao S., Yu S., Ping Ma L., Sun H. and Wang S. (2012). Fabrication of Fe_3O_4/SiO2 core/shell nanoparticles attached to graphene oxide and its use as an adsorbent. *Journal of Colloid and Interface Science*, **379**(1), 20–26.

Yip N. Y., Tiraferri A., Phillip W. A., Schiffman J. D. and Elimelech M. (2010). High performance thin-film composite forward osmosis membrane. *Environmental Science & Technology*, **44**(10), 3812–3818.

Zhang L., Zhong Y., Cha D. and Wang P. (2013). A self-cleaning underwater superoleophobic mesh for oil-water separation. *Scientific Reports*, **3**, 2326.

Zhang Y., Wei S., Liu F., Du Y., Liu S., Ji Y., Yokoi T., Tatsumi T. and Xiao F.-S. (2009). Superhydrophobic nanoporous polymers as efficient adsorbents for organic compounds. *Nano Today*, **4**(2), 135–142.

Zhang Z. and Kong J. (2011). Novel magnetic Fe_3O_4@C nanoparticles as adsorbents for removal of organic dyes from aqueous solution. *Journal of Hazardous Materials*, **193**, 325–329.

Zhang Z. and Wang P. (2012). Optimization of photoelectrochemical water splitting performance on hierarchical TiO_2 nanotube arrays. *Energy & Environmental Science*, **5**(4), 6506–6512.

Zhao S., Zou L., Tang C. Y. and Mulcahy D. (2012). Recent developments in forward osmosis: opportunities and challenges. *Journal of Membrane Science*, **396**, 1–21.

Zhao X. S. and Lu G. Q. (1998). Modification of MCM-41 by surface silylation with trimethylchlorosilane and adsorption study. *The Journal of Physical Chemistry B*, **102**(9), 1556–1561.

Zhu Y., Wang D., Jiang L. and Jin J. (2014). Recent progress in developing advanced membranes for emulsified oil/water separation. *NPG Asia Mater*, **6**, e101.

Chapter 2

Design and application of magnetic-core composite nano/micro particles for environmental remediation

Yuxiong Huang and Arturo A. Keller

2.1 INTRODUCTION

In the past decade, researchers' attention has been directed at the environmental applications of engineered nanomaterials (ENMs) due to the rapid growth of nanotechnology (Keller *et al.* 2013). ENMs' unique properties, such as large specific surface areas and high reactivities, make them excellent adsorbents (Huang & Keller, 2013), catalysts (Wang *et al.* 2014) and sensors (Ray, 2010) in various environmental applications (Adeleye *et al.* 2016).

Known as one of the cutting edge ENMs, magnetic particles (MP) are particularly attractive due to their superparamagnetic nature as well as their unique physicochemical properties such as high dispersibility, relative large surface area and the high ratio of surface to volume resulting in a higher adsorption capacity. Generally the core consists of magnetic elements such as iron, nickel, cobalt or their oxides and alloys, such that magnetic particles show ferromagnetic or superparamagnetic properties, which make them behave like small permanent magnets once magnetized as well as form lattice or aggregate due to magnetic interaction. Specifically, ferromagnetic particles have a permanent magnetism and removal of the magnetic field results in a lattice form, while superparamagnetic particles are attracted to a magnetic field but retain no residual magnetism after the field is removed. Superparamagnetic nano-adsorbents are particularly attractive as they can be easily retained and separated from treated water by applying external magnetic fields (as shown in Figure 2.1), which overcomes many of the issues present in filtration, centrifugation or gravitational separation, generally requiring less energy to achieve a given level of separation.

Iron oxides, such as magnetite (Fe_3O_4) and maghemite (γ-Fe_2O_3), are the most popular option owning to their small size and high surface area. However, pure iron oxide particles are prone to the formation of large aggregates due to the magnetic interaction, which results in changes of their magnetic properties, such as the loss of magnetism (Chen *et al.* 2011). In addition, their bare surface lacks selectivity, which eliminates their range of application and remediation capacity (Giakisikli & Anthemidis, 2013). Thus, researchers usually modify the surface of the iron oxide particles (as magnetic core) with specific functional groups and coatings forming a core-shell structure to overcome the above limitations (Giakisikli & Anthemidis, 2013). The coating onto MPs' surface, also known as a shell, can be obtained by the attachment or binding of inorganic components (e.g., silica or alumina, etc.) (Jiang *et al.* 2012; Karatapanis *et al.* 2011) or organic molecules (e.g., polymer or surfactant, etc.) (Faraji *et al.* 2010; Huang *et al.* 2010). With appropriate surface coating in this core-shell structure, it can help to improve their chemical stability (Ditsch *et al.* 2005), prevent their oxidation (Li *et al.* 2010) as well as lower their implication to the environment (Chen *et al.* 2011), and provide specific functionalities like selectivity for ion uptake (Koehler *et al.* 2009) or enhancing the water solubility of HOCs (Wang *et al.* 2009).

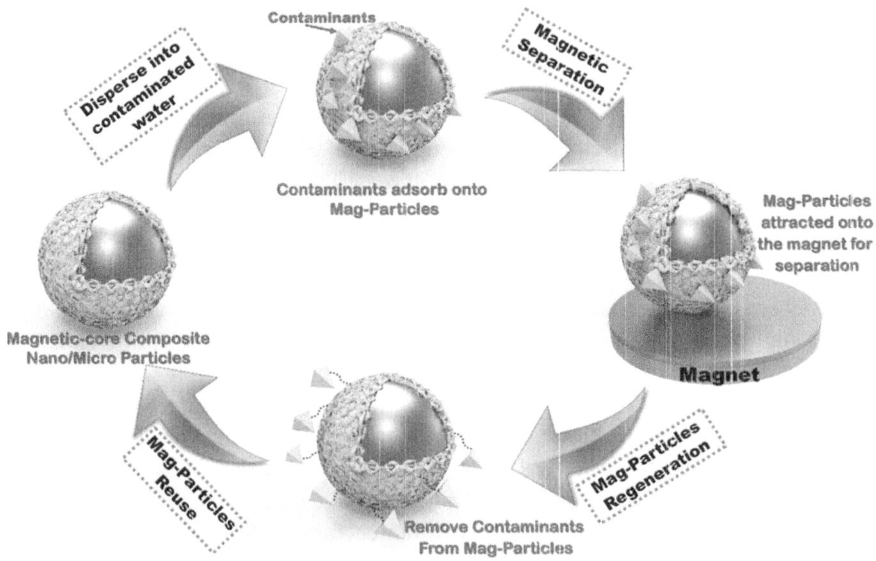

Figure 2.1 Procedure of applying magnetic-core composite nano/micro particles in water treatment.

2.2 SYNTHESIS OF MAGNETIC-CORE COMPOSITE NANO/MICRO PARTICLES

The synthesis of a magnetic-core composite nano/micro particles typically involves three steps (shown as Figure 2.2.):

1. Synthesis of magnetic particles, usually iron oxide (magnetite or maghemite) nanoparticles. Some studies (Huang & Keller, 2013) may use the commercial iron oxide nanoparticles to skip this step.
2. Coating of the magnetic core.
3. Surface modification of the coating layer.

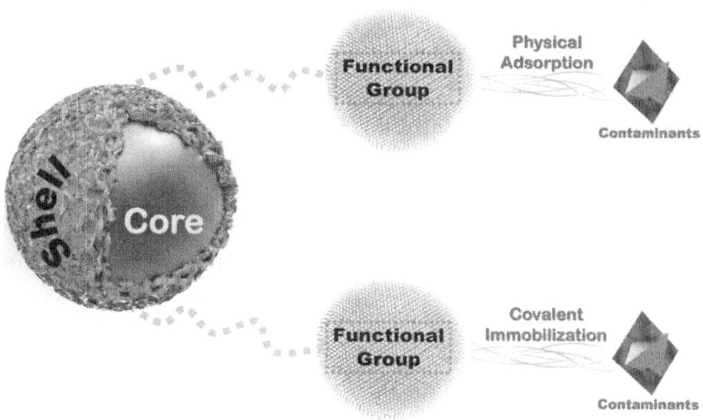

Figure 2.2 Functionalization of a core–shell type structure of magnetic-core composite nano/micro particles.

2.2.1 Synthesis of magnetic nanoparticles

There are many chemical methods that can be chosen to prepare magnetic nanoparticles, including classical chemical coprecipitation (Cornell & Schwertmann, 2003; Sugimoto, 2003; Schwarzer & Peukert, 2004), sol–gel syntheses (Dai *et al.* 2005; Duraes *et al.* 2005; Fouad *et al.* 2006), hydrothermal and high temperature reactions (Hyeon *et al.* 2001; Sun *et al.* 2004), surfactant mediate/template syntheses (in constrained environments) (Inouye *et al.* 1982; Deng *et al.* 2003), sonochemical reactions (Shafi *et al.* 1997; Koltypin *et al.* 1996), hydrolysis and thermolysis of precursors (Kimata *et al.* 2003), flow injection syntheses (Salazar-Alvarez *et al.* 2006), microemulsions (Solans *et al.* 2005; Geng *et al.* 2006), biomimetic mineralization (Allen *et al.* 2003; Rice *et al.* 2004), aerosol/vapor methods (Morales *et al.* 2003; Alexandrescu *et al.* 2005), and electrospray syntheses (Fürstner, 2008; Pascal *et al.* 1999). Among these methods, the chemical coprecipitation synthesis is one of the simplest and most efficient way for the preparation of magnetite particles (Laurent *et al.* 2010). In this method, iron oxides (Fe_3O_4 or γ-Fe_2O_3) are prepared by an aging stoichiometric mixture of ferrous and ferric salts in aqueous medium (generally involves the dissolution of a mixture of a solution of $FeCl_3{\cdot}6H_2O$ and $FeCl_2{\cdot}4H_2O$ in deionized water under nitrogen atmosphere with vigorous stirring at 70–85°C and the immediate addition of aqueous ammonia) (Jolivet *et al.* 2004).

2.2.2 Coating of magnetic core

Nowadays, there are various materials to coat the magnetic core, including silica (Huang & Keller, 2013; Wang *et al.* 2009), alumina (Tavallai, 2011), carbon (including carbon-based materials, e.g. activated carbon (Faraji *et al.* 2010), carbon nanotube (CNT) (Rastkari & Ahmadkhaniha, 2013) and graphene/graphene oxide (Liu *et al.* 2011)), polymer (Li *et al.* 2010), surfactant (Zhao *et al.* 2008) and biomass (e.g. pollen grains) (Thio *et al.* 2011). The mechanism of coating varies from covalent binding to electrostatic force, and the coating process also varies from sol–gel reaction to dissolution. For example, to conduct a silica coating onto magnetic core, the Stôber method (Stöber *et al.* 1968) through a sol–gel reaction (generally using an alkoxy silane (e.g. tetraethoxysilane) in acidic or basic media) is preferred. And during the coating process, the silane polymer can bind the magnetic iron oxide particles via a covalent bond (Yamaura *et al.* 2004), and enable the particles to have a strong affinity toward silica coating.

2.2.3 Surface modifications

To enhance remediation efficiency (Huang & Keller, 2013; Wang *et al.* 2009), the modifications with functional groups onto magnetic-core composite nano/micro particles' surface are usually needed. Typically, the surface modifications may involve the attachment of surfactants (Huang & Keller, 2013; Wang *et al.* 2009), cyclic oligosaccharides (e.g. β-Cyclodextrin (Ji *et al.* 2009)), functional groups (e.g. amine (Chen *et al.* 2013), thiol (Suleiman *et al.* 2009), carboxylic (Carpio *et al.* 2012) and C_{18} (Sha *et al.* 2008)), and chelators (e.g. EDTA (Koehler *et al.* 2009)).

2.3 TYPES OF MAGNETIC-CORE COMPOSITE NANO/MICRO PARTICLES

2.3.1 Silica-coated magnetic-core composite nano/micro particles

Silica is the most popular option for coating in terms of the mechanical and chemical stability under various environment (e.g. acidic conditions), high mass exchange as well as the high thermal resistance (Chen *et al.* 2011). Another advantage of a silica coating is that different functional groups can be attached onto the surface of silica-coated MPs by silanation using silane coupling agents (e.g. amine (Chen *et al.* 2013), thiol (Suleiman *et al.* 2009), carboxylic (Carpio *et al.* 2012) and C_{18} (Sha *et al.* 2008)) to enhance the selectivity and sorption capacity. A basic schematic diagram for the synthesis of silica-coated magnetic-core composite nano/micro particles modified with different functional groups is shown in Figure 2.3.

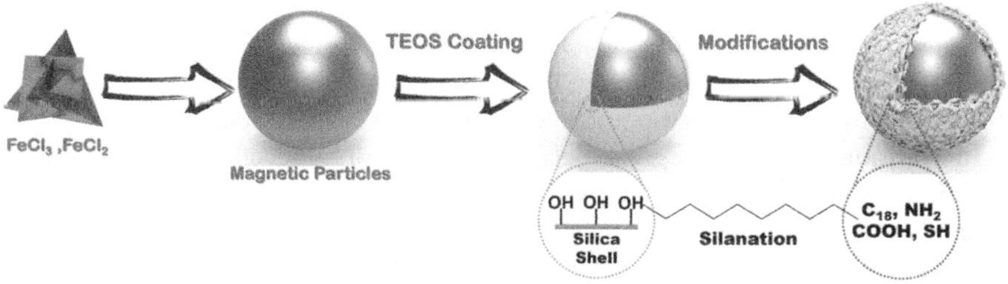

Figure 2.3 Schematic diagram of the synthesis of silica-coated magnetic-core composite nano/micro particles (MNPs) modified with different functional groups.

Recently magnetic permanently confined micelle arrays (Mag-PCMAs) were developed to treat hydrophobic organic contaminants (Wang *et al.* 2009). Mag-PCMAs are a novel composite material consisting of a mesoporous silica matrix, a co-deposited cationic surfactant micelle array, and a nano-scale magnetic silica iron oxide core. They have been applied to the removal of very hydrophobic compounds (Wang *et al.* 2009), pesticides (Clark & Keller, 2012), natural organic matter (Wang *et al.* 2011), oxyanions (Clark & Keller, 2012) and emerging organic contaminants (Huang & Keller, 2013). Mag-PCMAs can be recovered and reused many times, reporting nearly all (>95%) of the sorbed hydrophobic organic compounds (HOCs) could be recovered, with easy regeneration of Mag-PCMAs in five cycles of regeneration and reuse (Wang *et al.* 2009). The relatively low cost (~\$4/kg) (Huang & Keller, 2013) make it a sustainable approach for water treatment. In addition, researchers also studied the performance of different types of confined surfactants (cationic and nonionic surfactant) on the remediation of HOCs' remediation (Clark & Keller, 2012). The cationic surfactant (3-(trimethoxysily)propyl-octadecyldimethyl-ammonium chloride (TPODAC))-based Mag-PCMAs had better average recovery of the HOCs studied compared to the nonionic surfactant (Triton X-100)-based Mag-PCMAs, and was, in general, comparable to activated carbon. The porous structure of Mag-PCMAs can be optimized via a micelle swelling agent to achieve larger surface area and higher pore volume, to significantly improve the sorption kinetics and capacities of emerging organic compounds and PAHs.

Another type of superparamagnetic silica-coated ($Fe_3O_4@SiO_2$) magnetite nanoparticle modified by surfactants (cetylpyridinium chloride (CPC) or cetyltrimethylammonium bromide (CTAB)) has been synthesized and was applied as an effective sorbent material (>95% recovery rate) for several typical phenolic compounds (Zhao *et al.* 2008).

In a similar approach, a β-Cyclodextrin (β-CD) modified silica-coated magnetic sorbent was prepared (Ji *et al.* 2009). Due to the covalent bonding between $Fe_3O_4@SiO_2$ and β-CD, this sorbent showed high chemical stability and was used in a wide pH range (4.0–8.5), and was effective and rapid (~8 min) for sorbing bisphenol A (~97% recovery) and diethylstilbestrol (~95% recovery) from water.

Magnetic microsphere confined ionic liquid was synthesized by covalently bonding N-methylimidazolium on Fe_3O_4@SiO_2 microspheres (Yang *et al.* 2011), which werre used to extract three chlorophenols (CPs) from water showing good recoveries (70.3–88.8%) under optimized conditions (solution at pH 8 and 10 min mixing).

In addition to showing great performance on organic contaminants, silica-coated magnetic-core composite particles are also effective sorbents for inorganic contaminants, especially metal ions. For example, N-methyl-D-glucamine (NMDG) modified magnetic silica microparticles were prepared for the extraction of boron from aqueous solutions (Tural, 2010). The sorption and desorption efficiency of the microparticle sorbents were found to be 92.5 ± 0.5% and 99.8 ± 6%, respectively. Furthermore, the sorbent could be regenerated three times without considerably sorption capacity change.

An amino-functionalized Fe_3O_4@SiO_2 (i.e. silica shell on a magnetite core) magnetic nanomaterial was developed to remove metal ions from aqueous media (Wang *et al.* 2010). This nanoadsorbent showed high adsorption affinity for aqueous Cu(II), Pb(II), and Cd(II) ions, resulting from complexation of the metal ions by surface amino groups. Moreover, the adsorption affinity for heavy metal ions was not much impacted by the presence of humic acid or alkali/earth metal ions (Na^+, K^+, Mg^{2+}). The metal-loaded Fe_3O_4@SiO_2–NH_2 nanoparticles could be regenerated easily by acid treatment and reused for 4 cycles without significant change in sorption capacity.

Thiol-functionalized silica-coated magnetite nanoparticles (TF-SCMNPs) were synthesized for high efficiency removal and recovery of Hg(II) from water (Hakami *et al.* 2012). The particles have a uniform mesoporous structure with a pore size of ~2.1 nm, and the maximum adsorption capacity for Hg(II) was higher (up to 207.7 mg/g) than for other commonly used adsorbents, with 90% of Hg(II) removed during the first 5 min and equilibrium in less than 15 min (8 mg/L of TF-SCMNPs with 80 mg/L Hg(II) solution at pH 6.0). Hg(II) ions adsorbed on TF-SCMNPs can be desorbed effectively using thiourea in a 3 M HCl solution, and the TF-SCMNPs can be regenerated and maintained at almost the same Hg(II) adsorption capacity for 5 cycles.

Fe_3O_4@SiO_2 particles functionalized with organic moieties containing terminal dithiocarbamate groups were also used to remove Hg(II) from water (Girginova *et al.* 2010). The materials were reported to show high efficiency for Hg (II) uptake (74%) comparing to non-derivatized Fe_3O_4@SiO_2 (24%), which is due to the high stability of the chelates formed between the dithiocarbamate group and Hg(II). No information was provided on recovery and reuse potential.

A regenerable magnetic ligand particle (Mag-Ligand) which includes a metal-binding organic ligand (EDTA) attached to an iron oxide nanoparticle was developed for rapid remediation of Cd^{2+} and other metals from contaminated water (Huang & Keller, 2015). Mag-Ligand achieved above 97% Cd^{2+} removal under most environmental conditions (e.g. pH ranged from 4 to 10; a wide range concentration of Ca^{2+} (1–100 mg/L) or (0.6–60 mg/L) Mg^{2+} present in the solution) with relatively high sorption capacity (up to 141.5 mg/g) in less than 30 min, and was reused for 5 cycles with slight loss (within 10% change) of Cd^{2+} sorption capacity.

2.3.2 Magnetic-core composite nano/micro particles coated with other inorganic materials

Besides silica and carbon-based materials, there are other inorganic substance (e.g. hydrous aluminum oxide, manganese oxide) that have been used to coat magnetic-core composite nano/micro particles, to stabilize the magnetic particles under different environmental conditions (e.g. acidic solutions) and/or to provide high affinity binding sides to specific contaminants (Pommerenk & Schafran, 2005).

A magnetic nanosized adsorbent using hydrous aluminum oxide coated on a Fe_3O_4 nanoparticle (Fe_3O_4@$Al(OH)_3$ NPs), was prepared and applied to remove excessive fluoride from aqueous solution (Zhao *et al.* 2010). Due to the high affinity of hydrous aluminum oxide floc toward fluoride, Fe_3O_4@$Al(OH)_3$ NPs showed high adsorption capacity (up to 88.5 mg/g) at pH 6.5, and the residual concentration of fluoride using this adsorbent could be as low as 0.3 mg/L.

Hierarchical MnO_2-coated magnetic nanocomposite (Fe_3O_4@MnO_2) with a 3D flowerlike structure was synthesized by a mild hydrothermal process for removing heavy metal ions from contaminated water systems (Kim *et al.* 2013). The study indicated that Fe_3O_4@MnO_2 exhibited a greatly improved removal capacity toward four different heavy metals (Cd(II), Cu(II), Pb(II), and Zn(II)) compared to unmodified Fe_3O_4 nanoparticles (averagely the removal efficiency increased from ~20% to ~95%), and the maximum adsorption capacity toward Cd(II) was 53.2 mg/g. Fe_3O_4@MnO_2 retained over 80% of its adsorption capacity under various solution conditions and was affected by certain hydrochemical conditions, especially low pH, high ionic strength, and the presence of calcium ions. The nanocomposites can be easily recycled by using dilute acid and reused several times.

Ordered magnetic core–manganese oxide shell nanostructures were obtained by simply mixing manganese ferrite nanoparticles and potassium permanganate without any additional modified procedures (Zhai *et al.* 2009). This material can be used as absorbent for effectively removing organic dyes (e.g. sorption capacity of congo red was up to 42 mg/g), and can be regenerated at high temperature (300°C) while retaining a high removal capacity.

A nanosized core/shell structure containing a magnetite (Fe_3O_4) core and mesoporous ZrO_2 shell (Fe_3O_4@$mZrO_2$) with a relatively large surface area (107 m^2/g), synthesized using an inexpensive synthesis protocol, has been demonstrated as an efficient sorbent system for removal of phosphate ion (adsorption capacity was up to ~90%), and regeneration through the treatment with a KOH–KCl mixture allows Fe_3O_4@$mZrO_2$ reuse (Sarkar *et al.* 2010).

Magnetic TiO_2 composites (TiO_2@SiO_2@Fe_3O_4, TSM) were prepared using wet sol–gel and used to degrade dimethyl phthalate (DMP) via photocatalytic activity (Chang & Man, 2011). The study showed that the degradation of DMP using TSM followed pseudo-first-order kinetics (0.0014–0.0037 min^{-1}) and the adsorption capacity (~2–2.5 mg/g) and the photodegradation efficiency (~50%–65%) of DMP were significantly affected by the solution pH value, which both increased with decreasing pH.

Bentonite/iron oxide magnetic composites were prepared by co-precipitation and were used as a sorbent for the removal of Co(II) ions from radioactive wastewater (Chen *et al.* 2011). The results demonstrated that the sorption of Co(II) was strongly dependent on pH and ionic strength at low pH values with a 13.68 mg/g maximum sorption capacities at pH 5.0 and increased to 22.73 mg/g when at pH 6.6.

2.3.3 Carbon-coated magnetic-core composite nano/micro particles

Carbon-based nanomaterials (e.g. activated carbon, carbon nanotubes (CNTs), graphene/graphene oxide) can exhibit high chemical and thermal stability and biocompatibility, and the possibility of surface modification and additional porosity make them an interesting coating material for magnetic particles (Xie *et al.* 2014).

Activated carbon/iron oxide magnetic composites (~100 mesh) were prepared to be used as adsorbent for a wide range of contaminants including volatile organic compounds (VOCs) such as chloroform, phenol, chlorobenzene and drimaren red dye from aqueous solution (Oliveira *et al.* 2002). The study indicated that there was no significant decrease in the surface area or the porosity of the activated carbon. In addition, the sorption capacity was unchanged even with the presence of the iron oxides in the composite (sorption capacity of magnetic composites for phenol, chlorobenzene and chloroform was 117, 305 and 710 mg/g, while the pure activated carbon was 162, 480 and 910, respectively), and the magnetic composites showed good resistance over a pH range of 5–11.

In another study, multiwall carbon nanotubes (MWCNTs)/nano-iron oxide magnetic composites were prepared. This composite showed good adsorption capacity for chromium ions (Gupta *et al.* 2011). The researchers demonstrated that adsorption capability of the composite (~18 mg/g) is higher than that of MWCNTs (~12 mg/g) and activated carbon (~1.6 mg/g), which is increased by increasing the agitation speed and it is pH dependent in the batch adsorption while is increased by decreasing the flow rate in the continuous flow sorption.

A graphene nanosheet (GNS)/magnetite (Fe_3O_4) composite was synthesized by one-step solvothermal method for fast adsorption of an organic dye, methylene blue (MB), from water (Ai *et al.* 2011). Due to the electrostatic attraction between the negative surface oxygen-containing groups and cationic MB and the π–π interactions between the MB molecules and the aromatic rings of graphene, the GNS/Fe_3O_4 composite exhibited extraordinary removal capacity (up to 43.82 mg/g), without significant decrease compared to pure GNS (~60 mg/g).

Hybrids of graphene oxide (GO) nanosheets and surface-modified Fe_3O_4 NPs were fabricated by a two-step process, and the covalent bonding between Fe_3O_4 NPs and GO was realized for the first time (He *et al.* 2010). The adsorption capacity of GO–Fe_3O_4 for MB and neutral red cationic dyes was as high as 190.1 and 140.8 mg/g, respectively.

2.3.4 Polymer coated magnetic-core composite nano/micro particles

Polymers can be chemically or physically anchored on magnetic particles by covalent bonding or by electrostatic interaction (Xie *et al.* 2014). Using different polymeric coating (natural or synthetic polymer), these magnetic-core composite nano/micro particles can be tailored to adsorb different pollutants.

Magnetic polyaniline (PANI) polymer nanocomposites (PNCs) were prepared by using a surface initiated polymerization (SIP) method and were applied to remove Cr(VI) from polluted solutions within a wide pH range (pH 1–11) (Gu *et al.* 2012). The results indicated that PANI was partially oxidized after treatment with 4 mg/L Cr(VI) solution, with Cr(VI) being reduced to Cr(III). PNCs could rapidly and effectively treat Cr(VI) (e.g. by using 0.6 g/L PNCs can completely remove 20.0 mL Cr(VI) neutral solution (initial concentration = 1.0–3.0 mg/L) after 5 min treatment) contaminated water. The PANI-PNC can be regenerated by being mixed by either 1 mol/L *p*-toluene sulfonic acid (PTSA) or 1 mol/L hydrochloric acid (HCl) for 5 min under sonication.

Poly(1-vinylimidazole) with a trimethoxysilyl terminal group (Im_{18}) was grafted onto magnetic nanoparticles via siloxane bonds to produce polymer-grafted magnetic nanocomposite particles (Mag-Im_{18}), which were chemically stable over a pH range of 3.5–10.0 (Takafuji *et al.* 2004). Mag-Im_{18} can be used for metal ions removal and showed a selective binding of divalent metal ions with a binding strength following $Cu^{2+} \gg Ni^{2+} > Co^{2+}$.

Polyethylene imine (PEI) coated nanomagnets with ethylenediaminetetraacetic acid (EDTA) like chelators modification were prepared for fast removal of Cd, Cu and Pb from waste water (Koehler *et al.* 2009). The sorbents showed high removal efficiency on heavy metal ions at different concentrations ranging from 10 mg/L to 1 μg/L, and can be used numerous times.

A copolymer of methyl methacrylate (MMA) and maleic anhydride (MA), poly(MMA-*co*-MA), Poly(MMA-*co*-MA) modified Fe_3O_4 nanoparticles were synthesized for removal of metal ions such as Cd^{2+}, Cr^{3+}, Co^{2+}, and Zn^{2+} from aqueous solution (Masoumi *et al.* 2014). Heavy metal ions' sorption onto the Poly(MMA-*co*-MA)@Fe_3O_4 was rapid (~1h) and effective with the maximum adsorption capacity of 90.09, 90.91, 109.89, and 111.11 mg g^{-1} being found for $Co^{2+} < Cr^{3+} < Zn^{2+} < Cd^{2+}$, respectively.

Molecular imprinted polymers (MIPs) are a type of synthesized material with specific recognition ability for the template molecules and high reusability, selectivity and low consumption (Le Noir *et al.* 2006). Novel magnetic and hydrophilic molecularly imprinted polymers (mag-MIPs) were prepared by an inverse emulsion–suspension polymerization to remove water-soluble acid dyes from contaminated water with 1-(α-methyl acrylate)-3-methylimidazolium bromide (1-MA-3MI-Br) being utilized as a new functional monomer (Luo *et al.* 2011). According to the study, the removal efficiency toward water-soluble acid dyes is very high with all above 95% in distilled water, tap water, river water and wastewater with an efficient way to recycle the water-soluble acid dyes in water media and can be reused at least five times without obvious decrease in the removal efficiency.

2.3.5 Surfactant coated magnetic-core composite nano/micro particles

Absorption of ionic surfactants (e.g. sodium dodecylesulfate (SDS), cetyltrimethylammonium bromide (CTAB)) on mineral oxides (e.g., alumina, silica, titanium dioxide, and ferric oxyhydroxides) can lead to the formation of hemimicelles (monolayers of surfactants attached on an oppositely charged mineral-oxide surface with the hydrocarbon tail groups protruding into the solution) and admicelles (formed by hydrophobic interactions between the tails of the surfactant hydrocarbon chains when surfactant saturated). Since the outer surface of hemimicelles is hydrophobic whereas that of admicelles is ionic, it provides different mechanisms (e.g., hydrophobic interactions, electrostatic interactions) for interaction with pollutant compounds.

SDS-coated Fe_3O_4 NPs as an efficient adsorbent was prepared to remove the safranin O dye from wastewater samples (Shariati *et al.* 2011). Due to their very high surface areas, high adsorption capacity (up to 769.23 mg/g) can be achieved in short times and can be simply regenerated for several (~8) cycles reuse.

CTAB modified magnetic nanoparticles (Fe_3O_4@CTAB) were synthesized by a modified simple co-precipitation process and used to remove arsenate from water (Jin *et al.* 2012). Adsorption of arsenate on Fe_3O_4@CTAB reached equilibrium within 2 min at pH 6, and the adsorption capacity was up to 23.07 mg/g (3 times as that of bare Fe_3O_4). The fabricated Fe_3O_4@CTAB could be used under different environmental conditions including across wide range pH (from 3 to 9), with the presence of NOM (humic acid or alginate) in solutions (from 1 to 10 mg L^{-1} as TOC), with presence of sulfate, bicarbonate, and silicate (up to 20 mg L^{-1}). Furthermore, Fe_3O_4@CTAB could be regenerated with a simple alkalization-acidification process and higher than 85% As(V) removal rates can be achieved even in the fifth adsorption/desorption cycle.

The adsorption and removal of amaranth (AM) from an aqueous solution by Fe_3O_4@CTAB as adsorbent was reported (Zargar *et al.* 2009). The adsorbent showed high adsorption capacity (1.05 mg/mg) and fast adsorption and desorption rates (<3 min) for the removal of AM due to the high specific surface area and the absence of internal diffusion resistance, and the researcher suggested that it could be employed in the removal of the anionic dye from wastewater.

2.3.6 Other organic materials coated/functionalized magnetic-core composite nano/micro particles

There are various organic materials such as dimercaptosuccinic acid (DMSA), EDTA, chitosan, and calcium alginate, etc., which can form a coating shell or surface functionalization for magnetic-core composite nano/micro particles.

Iron oxide (Fe_3O_4) nanoparticles were functionalized with DMSA to generate a highly dispersible sorbent, stable in solutions and witha large surface area (114 m^2/g), and high functional group content (1.8 mmol thiols/g) to remove metals such as Hg, Ag, Pb, Cd, and Tl (Yantasee *et al.* 2007). DMSA-Fe_3O_4 showed higher sorption capacity (~227 mg of Hg/g) and faster rate (removed 99% of 1 mg/L Pb within a minute) when compared to other commercial sorbents.

Warner *et al.* reported functionalization of superparamagnetic iron oxide nanoparticles using a straightforward precursor synthesis followed by a facile ligand exchange process to bind readily available affinity ligands onto the nanoparticle surface which allowed for tailoring of the surface chemistry to impart the specificity and affinity toward the target analytes (heavy metals of environmental concern) (Warner *et al.* 2010). In the study, Thiol and EDTA surface chemistries (including lauric acid (LA), EDTA, l-glutathione (GSH), mercaptobutyric acid (MBA), α-thio-ω-(propionic acid) hepta(ethylene glycol) (PEG-SH), and *meso*-2,3-dimercaptosuccinic acid (DMSA)) were added to Fe_3O_4 nanoparticles and were shown to be highly effective, magnetically active sorbent materials for heavy metal ions such as Hg, Pb, Cd, Ag, Co, Cu, and Tl.

Chitosan-coated magnetic nanoparticles (CCMNPs), modified with a biodegradable and eco-friendly biologic reagent, α-ketoglutaric acid (α-KA), was used as a magnetic nanoadsorbent to remove toxic Cu^{2+} ions from aqueous solution (Zhou *et al.* 2009). Showing high stability and recovery capacities, adsorption of Cu^{2+} onto CCMNPs reaches equilibrium within 60 min; 50% of Cu^{2+} was adsorbed after 2 min with a relatively large adsorption capacity (up to 96.15 mg/g).

Via a coprecipitation procedure, humic acid (HA) coated Fe_3O_4 nanoparticles (Fe_3O_4@HA) were developed for the removal of toxic Hg(II), Pb(II), Cd(II), and Cu(II) from water (Liu *et al.* 2008). Sorption of the heavy metals to Fe_3O_4@HA reached equilibrium in less than 15 min with maximum adsorption capacities from 46.3 to 97.7 mg/g and over 95%–99% removal efficiency. The Fe_3O_4/HA was stable in tap water, natural waters, and acidic/basic solutions ranging from 0.1 M HCl to 2 M NaOH with low leaching of Fe (≤3.7%) and HA (≤5.3%).

A functionalizable organosiliceous hybrid magnetic material (Fe_3O_4@POSS-SH) was easily constructed by surface polymerization of octavinyl polyhedral oligomeric silsesquioxane (POSS) on the Fe_3O_4 nanoparticles followed by being tethered with an organic chain containing dithiol via thiol–ene addition reaction (He *et al.* 2013). It was demonstrated to be a convenient, efficient single adsorbent for simultaneous elimination of inorganic heavy metal ions and organic dyes in simulate multicomponent wastewater at ambient temperaturea and can be easily regenerated with either methanol–acetic acid (for organic dyes) or hydrochloric acid (for heavy metal ions) under ultrasonication.

A calcium-alginate magnetic micro particles (309.6 μm) sorbent with a specific surface area 312.94 m^2/g, was prepared by an electrostatic extrusion technique for arsenic(V) and copper remediation (Lim & Chen, 2007). According to the study, the maximum adsorption capacities of arsenic and copper ions are 6.75 and 60.24 mg/g, respectively, and the adsorption was affected by the solution pH.

2.3.7 Magnetized biomass composite nano/micro particles

Biomass materials are usually available in large quantities, sometimes as waste products from industrial or agricultural operations, and they have the potential to be low-cost biosorbents due to the presence of various functional groups on their cell wall, which offer certain forces of attractions for the metal ions or organic compounds and provide a high removal efficency (Farooq *et al.* 2010).

Magnetized pollen grains (MRP) were prepared to be used as effective sorbents for HOCs (including acenaphthene, phenanthrene, atrazine, diuron, and lindane) in contaminated water and with similar levels of organic contaminants removal as traditional activated carbon (Thio *et al.* 2011). Furthermore, the MRP could be regenerated and reused using common organic solvents to extract the HOCs.

A magnetic derivative of spent grain was prepared by contact with water-based magnetic fluid and used for organic dye removal with high adsorption capacity (up to 72.4 mg of dye (Bismarck brown Y) per g of dried magnetically modified spent grain) (Safarik *et al.* 2011). Through similar magnetically modification, a new type of magnetically responsive Fodder yeast (*Kluyveromyces fragilis*) cells biocomposite material for the removal of water-soluble dyes with relatively high adsorption capacity (ranged between 29.9 (amido black 10B) and 138.2 (safranin O) mg of dye per g of dried magnetically modified cells) (Safarik *et al.* 2007).

2.4 CONCLUSIONS

The design and application of magnetic-core composite nano/micro particles for environmental remediation has been an interesting topic over the last few years, and there is a clearly increasing trend in the number of works being published in this field. The magnetic-core composite nano/micro particles has combined the advantages of magnetic particles and modified layers in excellent sorbents for environmental remediation:

1. High removal efficiency due to the high surface area to volume ration
2. Rapid and sustainable magnetic separation after treatment
3. High adsorption capacity with various modified layers to certain pollutants
4. Good reusability and easy regeneration

2.5 REFERENCES

Adeleye A. S., Conway J. R., Garner K., Huang Y., Su Y. and Keller A. A. (2016). Engineered nanomaterials for water treatment and remediation: Costs, benefits, and applicability. *Chemical Engineering Journal*, **286**, 640–662.

Ai L. H., Zhang C. Y. and Chen Z. L. (2011). Removal of methylene blue from aqueous solution by a solvothermal-synthesized graphene/magnetite composite. *Journal of Hazardous Materials*, **192**(3), 1515–1524.

Alexandrescu R., Morjan I., Voicu I., Dumitrache F., Albu L., Soare I. and Prodan G. (2005). Combining resonant/non-resonant processes: Nanometer-scale iron-based material preparation via CO_2 laser pyrolysis. *Applied Surface Science*, **248**(1), 138–146.

Allen M., Willits D., Young M. and Douglas T. (2003). Constrained synthesis of cobalt oxide nanomaterials in the 12-subunit protein cage from Listeria innocua. *Inorganic Chemistry*, **42**(20), 6300–6305.

Carpio A., Mercader-Trejo F., Arce L. and Valcarcel M. (2012). Use of carboxylic group functionalized magnetic nanoparticles for the preconcentration of metals in juice samples prior to the determination by capillary electrophoresis. *Electrophoresis*, **33**(15), 2446–2453.

Chang C. F. and Man C. Y. (2011). Titania-coated magnetic composites as photocatalysts for phthalate photodegradation. *Industrial & Engineering Chemistry Research*, **50**(20), 11620–11627.

Chen B., Hu B., He M., Huang Q., Zhang Y. and Zhang X. (2013). Speciation of selenium in cells by HPLC-ICP-MS after (on-chip) magnetic solid phase extraction. *Journal of Analytical Atomic Spectrometry*, **28**(3), 334–343.

Chen J. W., Xiu Z. M., Lowry G. V. and Alvarez P. J. J. (2011). Effect of natural organic matter on toxicity and reactivity of nano-scale zero-valent iron. *Water Research*, **45**(5), 1995–2001.

Chen L., Huang Y., Huang L. L., Liu B., Wang G. and Yu S. M. (2011). Characterization of Co(II) removal from aqueous solution using bentonite/iron oxide magnetic composites. *Journal of Radioanalytical and Nuclear Chemistry*, **290**(3), 675–684.

Chen L. G., Wang T. and Tong J. (2011). Application of derivatized magnetic materials to the separation and the preconcentration of pollutants in water samples. *TrAC Trends in Analytical Chemistry*, **30**(7), 1095–1108.

Clark K. K. and Keller A. A. (2012a). Adsorption of perchlorate and other oxyanions onto magnetic permanently confined micelle arrays (Mag-PCMAs). *Water Research*, **46**(3), 635–644.

Clark K. K. and Keller A. A. (2012b). Investigation of two magnetic permanently confined micelle array sorbents using nonionic and cationic surfactants for the removal of PAHs and pesticides from aqueous media. *Water, Air, & Soil Pollution*, **223**(7), 3647–3655.

Cornell R. M. and Schwertmann U. (2003). The Iron Oxides: Structure, Properties, Reactions, Occurrences and Uses. John Wiley & Sons (Wiley VCH), Weinheim, Germany.

Dai Z., Meiser F. and Möhwald H. (2005). Nanoengineering of iron oxide and iron oxide/silica hollow spheres by sequential layering combined with a sol–gel process. *Journal of Colloid and Interface Science*, **288**(1), 298–300.

Deng Y., Wang L., Yang W., Fu S. and Elaıssari A. (2003). Preparation of magnetic polymeric particles via inverse microemulsion polymerization process. *Journal of Magnetism and Magnetic Materials*, **257**(1), 69–78.

Ditsch A., Laibinis P. E., Wang D. I. C. and Hatton T. A. (2005). Controlled clustering and enhanced stability of polymer-coated magnetic nanoparticles. *Langmuir*, **21**(13), 6006–6018.

Duraes L., Costa B., Vasques J., Campos J. and Portugal A. (2005). Phase investigation of as-prepared iron oxide/hydroxide produced by sol–gel synthesis. *Materials Letters*, **59**(7), 859–863.

Faraji M., Yamini Y. and Rezaee M. (2010a). Extraction of trace amounts of mercury with sodium dodecyle sulphate-coated magnetite nanoparticles and its determination by flow injection inductively coupled plasma-optical emission spectrometry. *Talanta*, **81**(3), 831–836.

Faraji M., Yamini Y., Saleh A., Rezaee M., Ghambarian M. and Hassani R. (2010b). A nanoparticle-based solid-phase extraction procedure followed by flow injection inductively coupled plasma-optical emission spectrometry to determine some heavy metal ions in water samples. *Analytica Chimica Acta*, **659**(1–2), 172–177.

Farooq U., Kozinski J. A., Khan M. A. and Athar M. (2010). Biosorption of heavy metal ions using wheat based biosorbents – A review of the recent literature. *Bioresource Technology*, **101**(14), 5043–5053.

Fouad O., Ismail A., Zaki Z. and Mohamed R. (2006). Zinc oxide thin films prepared by thermal evaporation deposition and its photocatalytic activity. *Applied Catalysis B: Environmental*, **62**(1), 144–149.

Fürstner A. (2008). Active metals: Preparation, characterization, applications, John Wiley & Sons (Wiley VCH), Weinheim, Germany.

Geng F., Zhao Z., Cong H., Geng J. and Cheng H.-M. (2006). An environment-friendly microemulsion approach to α-FeOOH nanorods at room temperature. *Materials Research Bulletin*, **41**(12), 2238–2243.

Giakisikli G. and Anthemidis A. N. (2013). Magnetic materials as sorbents for metal/metalloid preconcentration and/or separation. A review. *Analytica Chimica Acta*, **789**, 1–16.

Girginova P. I., Daniel-Da-Silva A. L., Lopes C. B., Figueira P., Otero M., Amaral V. S., Pereira E. and Trindade T. (2010). Silica coated magnetite particles for magnetic removal of Hg2+ from water. *Journal of Colloid and Interface Science*, **345**(2), 234–240.

Gu H. B., Rapole S. B., Sharma J., Huang Y. D., Cao D. M., Colorado H. A., Luo Z. P., Haldolaarachchige N., Young D. P., Walters B., Wei S. Y. and Guo Z. H. (2012). Magnetic polyaniline nanocomposites toward toxic hexavalent chromium removal. *RSC Advances*, **2**(29), 11007–11018.

Gupta V. K., Agarwal S. and Saleh T. A. (2011). Chromium removal by combining the magnetic properties of iron oxide with adsorption properties of carbon nanotubes. *Water Research*, **45**(6), 2207–2212.

Hakami O., Zhang Y. and Banks C. J. (2012). Thiol-functionalised mesoporous silica-coated magnetite nanoparticles for high efficiency removal and recovery of Hg from water. *Water Research*, **46**(12), 3913–3922.

He F. A., Fan J. T., Ma D., Zhang L. M., Leung C. and Chan H. L. (2010). The attachment of Fe3O4 nanoparticles to graphene oxide by covalent bonding. *Carbon*, **48**(11), 3139–3144.

He H. B., Li B., Dong J. P., Lei Y. Y., Wang T. L., Yu Q. W., Feng Y. Q. and Sun Y. B. (2013). Mesostructured Nanomagnetic Polyhedral Oligomeric Silsesquioxanes (POSS) incorporated with dithiol organic anchors for multiple pollutants capturing in wastewater. *ACS Applied Materials & Interfaces*, **5**(16), 8058–8066.

Huang Y. and Keller A. A. (2015). EDTA functionalized magnetic nanoparticle sorbents for cadmium and lead contaminated water treatment. *Water Research*, **80**, 159–168.

Huang Y. F., Li Y., Jiang Y. and Yan X. P. (2010). Magnetic immobilization of amine-functionalized magnetite microspheres in a knotted reactor for on-line solid-phase extraction coupled with ICP-MS for speciation analysis of trace chromium. *Journal of Analytical Atomic Spectrometry*, **25**(9), 1467–1474.

Huang Y., Fulton A. N. and Keller A. A. (2016). Optimization of porous structure of superparamagnetic nanoparticle adsorbents for higher and faster removal of emerging organic contaminants and PAHs. *Environmental Science: Water Research & Technology*. doi: 10.1039/C6EW00066E.

Huang Y. and Keller A. A. (2013). Magnetic nanoparticle adsorbents for emerging organic contaminants. *ACS Sustainable Chemistry & Engineering*, **1**(7), 731–736.

Hyeon T., Lee S. S., Park J., Chung Y. and Na H. B. (2001). Synthesis of highly crystalline and monodisperse maghemite nanocrystallites without a size-selection process. *Journal of the American Chemical Society*, **123**(51), 12798–12801.

Inouye K., Endo R., Otsuka Y., Miyashiro K., Kaneko K. and Ishikawa T. (1982). Oxygenation of ferrous ions in reversed micelle and reversed microemulsion. *The Journal of Physical Chemistry*, **86**(8), 1465–1469.

Ji Y. S., Liu X. Y., Guan M., Zha C. D., Huang H. Y., Zhang H. X. and Wang C. M. (2009). Preparation of functionalized magnetic nanoparticulate sorbents for rapid extraction of biphenolic pollutants from environmental samples. *Journal of Separation Science*, **32**(12), 2139–2145.

Jiang H. M.; Yan Z. P., Zhao Y., Hu X. and Lian H. Z. (2012). Zincon-immobilized silica-coated magnetic Fe3O4 nanoparticles for solid-phase extraction and determination of trace lead in natural and drinking waters by graphite furnace atomic absorption spectrometry. *Talanta*, **94**, 251–256.

Jin Y. J., Liu F., Tong M. P. and Hou Y. L. (2012). Removal of arsenate by cetyltrimethylammonium bromide modified magnetic nanoparticles. *Journal of Hazardous Materials*, **227**, 461–468.

Jolivet J.-P., Chanéac C. and Tronc E. (2004). Iron oxide chemistry. From molecular clusters to extended solid networks. *Chemical Communications* **5**, 481–483.

Karatapanis A. E., Fiamegos Y. and Stalikas C. D. (2011). Silica-modified magnetic nanoparticles functionalized with cetylpyridinium bromide for the preconcentration of metals after complexation with 8-hydroxyquinoline. *Talanta*, **84**(3), 834–839.

Keller A. A., McFerran S., Lazareva A. and Suh S. (2013). Global life cycle releases of engineered nanomaterials. *Journal of Nanoparticle Research*, **15**(6), 1–17.

Kim E. J., Lee C. S., Chang Y. Y. and Chang Y. S. (2013). Hierarchically Structured Manganese Oxide-Coated Magnetic Nanocomposites for the Efficient Removal of Heavy Metal Ions from Aqueous Systems. *ACS Applied Materials & Interfaces*, **5**(19), 9628–9634.

Kimata M., Nakagawa D. and Hasegawa M. (2003). Preparation of monodisperse magnetic particles by hydrolysis of iron alkoxide. *Powder Technology*, **132**(2), 112–118.

Koehler F. M., Rossier M., Waelle M., Athanassiou E. K., Limbach L. K., Grass R. N., Gunther D. and Stark W. J. (2009). Magnetic EDTA: coupling heavy metal chelators to metal nanomagnets for rapid removal of cadmium, lead and copper from contaminated water. *Chemical Communications*, 32, 4862–4864.

Koltypin Y., Katabi G., Cao X., Prozorov R. and Gedanken A. (1996). Sonochemical preparation of amorphous nickel. *Journal of Non-crystalline Solids*, **201**(1), 159–162.

Laurent S., Forge D., Port M., Roch A., Robic C., Elst L. V. and Muller R. N. (2010). Magnetic Iron Oxide Nanoparticles: Synthesis, Stabilization, Vectorization, Physicochemical Characterizations, and Biological Applications (vol 108, pg 2064, 2008). *Chemical Review*, **110**(4), 2574–2574.

Le Noir M., Guieysse B. and Mattiasson B. (2006). Removal of trace contaminants using molecularly imprinted polymers. *Water Science and Technology*, **53**(11), 205–212.

Li Q. L., Lam M. H. W., Wu R. S. S. and Jiang B. W. (2010). Rapid magnetic-mediated solid-phase extraction and pre-concentration of selected endocrine disrupting chemicals in natural waters by poly(divinylbenzene-co-methacrylic acid) coated Fe_3O_4 core-shell magnetite microspheres for their liquid chromatography-tandem mass spectrometry determination. *Journal of Chromatography. A*, **1217**(8), 1219–1226.

Li Z. Q., Greden K., Alvarez P. J. J., Gregory K. B. and Lowry G. V. (2010). Adsorbed Polymer and NOM Limits Adhesion and Toxicity of Nano Scale Zerovalent Iron to *E. coli*. *Environmental Science and Technology*, **44**(9), 3462–3467.

Lim S. F. and Chen J. P. (2007). Synthesis of an innovative calcium-alginate magnetic sorbent for removal of multiple contaminants. *Applied Surface Science*, **253**(13), 5772–5775.

Liu J. F., Zhao Z. S. and Jiang G. B. (2008). Coating Fe_3O_4 magnetic nanoparticles with humic acid for high efficient removal of heavy metals in water. *Environmental Science and Technology*, **42**(18), 6949–6954.

Liu M. C., Chen C. L., Hu J., Wu X. L. and Wang X. K. (2011). Synthesis of Magnetite/Graphene Oxide Composite and Application for Cobalt(II) Removal. *Journal of Physical Chemistry C*, **115**(51), 25234–25240.

Luo X. B., Zhan Y. C., Huang Y. N., Yang L. X., Tu X. M. and Luo S. L. (2011). Removal of water-soluble acid dyes from water environment using a novel magnetic molecularly imprinted polymer. *Journal of Hazardous Materials*, **187**(1–3), 274–282.

Masoumi A., Ghaemy M. and Bakht A. N. (2014). Removal of Metal Ions from Water Using Poly(MMA-co-MA)/Modified-Fe_3O_4 Magnetic Nanocomposite: Isotherm and Kinetic Study. *Industrial & Engineering Chemistry Research*, **53**(19), 8188–8197.

Morales M., Bomati-Miguel O., Pérez de Alejo R., Ruiz-Cabello J., Veintemillas-Verdaguer S. and O'Grady K. (2003). Contrast agents for MRI based on iron oxide nanoparticles prepared by laser pyrolysis. *Journal of Magnetism and Magnetic Materials*, **266**(1), 102–109.

Oliveira L. C. A., Rios R. V. R. A., Fabris J. D., Garg V., Sapag K., Lago R. M. (2002). Activated carbon/iron oxide magnetic composites for the adsorption of contaminants in water. *Carbon*, **40**(12), 2177–2183.

Pascal C., Pascal J., Favier F., Elidrissi Moubtassim M. and Payen C. (1999). Electrochemical synthesis for the control of γ-Fe_2O_3 nanoparticle size. Morphology, microstructure, and magnetic behavior. *Chemistry of Materials*, **11**(1), 141–147.

Pommerenk P. and Schafran G. C. (2005). Adsorption of inorganic and organic ligands onto hydrous aluminum oxide: Evaluation of surface charge and the impacts on particle and NOM removal during water treatment. *Environmental Science and Technology*, **39**(17), 6429–6434.

Rastkari N. and Ahmadkhaniha R. (2013). Magnetic solid-phase extraction based on magnetic multi-walled carbon nanotubes for the determination of phthalate monoesters in urine samples. *Journal of Chromatography. A*, **1286**, 22–28.

Ray P. C. (2010). Size and Shape Dependent Second Order Nonlinear Optical Properties of Nanomaterials and Their Application in Biological and Chemical Sensing. *Chemical Review*, **110**(9), 5332–5365.

Rice G., Tang L., Stedman K., Roberto F., Spuhler J., Gillitzer E., Johnson J. E., Douglas T. and Young M. (2004). The structure of a thermophilic archaeal virus shows a double-stranded DNA viral capsid type that spans all domains of life. *Proceedings of the National Academy of Sciences of the United States of America*, **101**(20), 7716–7720.

Safarik I., Rego L. F. T., Borovska M., Mosiniewicz-Szablewska E., Weyda F. and Safarikova M. (2007). New magnetically responsive yeast-based biosorbent for the efficient removal of water-soluble dyes. *Enzyme and Microbial Technology*, **40**(6), 1551–1556.

Safarik I., Horska K. and Safarikova M. (2011). Magnetically modified spent grain for dye removal. *Journal of Cereal Science*, **53**(1), 78–80.

Salazar-Alvarez G., Muhammed M. and Zagorodni A. A. (2006). Novel flow injection synthesis of iron oxide nanoparticles with narrow size distribution. *Chemical Engineering Science*, **61**(14), 4625–4633.

Sarkar A., Biswas S. K. and Pramanik P. (2010). Design of a new nanostructure comprising mesoporous ZrO_2 shell and magnetite core (Fe_3O_4@ $mZrO_2$) and study of its phosphate ion separation efficiency. *Journal of Materials Chemistry*, **20**(21), 4417–4424.

Schwarzer H.-C. and Peukert W. (2004). Tailoring particle size through nanoparticle precipitation. *Chemical Engineering Communications*, **191**(4), 580–606.

Sha Y., Deng C. and Liu B. (2008). Development of C_{18}-functionalized magnetic silica nanoparticles as sample preparation technique for the determination of ergosterol in cigarettes by microwave-assisted derivatization and gas chromatography/mass spectrometry. *Journal of Chromatography. A*, **1198**, 27–33.

Shafi K. V., Koltypin Y., Gedanken A., Prozorov R., Balogh J., Lendvai J. and Felner I. (1997). Sonochemical preparation of nanosized amorphous $NiFe_2O_4$ particles. *The Journal of Physical Chemistry B*, **101**(33), 6409–6414.

Shariati S., Faraji M., Yamini Y. and Rajabi A. A. (2011). Fe_3O_4 magnetic nanoparticles modified with sodium dodecyl sulfate for removal of safranin O dye from aqueous solutions. *Desalination*, **270**(1–3), 160–165.

Solans C., Izquierdo P., Nolla J., Azemar N. and Garcia-Celma M. (2005). Nano-emulsions. *Current Opinion in Colloid & Interface Science*, **10**(3), 102–110.

Stöber W., Fink A. and Bohn E. (1968). Controlled growth of monodisperse silica spheres in the micron size range. *Journal of Colloid and Interface Science*, **26**(1), 62–69.

Sugimoto T. (2003). Formation of Monodispersed Nano-and Micro-Particles Controlled in Size, Shape, and Internal Structure. *Chemical Engineering & Technology*, **26**(3), 313–321.

Suleiman J. S., Hu B., Peng H. Y. and Huang C. Z. (2009). Separation/preconcentration of trace amounts of Cr, Cu and Pb in environmental samples by magnetic solid-phase extraction with Bismuthiol-II-immobilized magnetic nanoparticles and their determination by ICP-OES. *Talanta*, **77**(5), 1579–1583.

Sun S., Zeng H., Robinson D. B., Raoux S., Rice P. M., Wang S. X. and Li G. (2004). Monodisperse MFe_2O_4 (M = Fe, Co, Mn) nanoparticles. *Journal of the American Chemical Society*, **126**(1), 273–279.

Takafuji M., Ide S., Ihara H. and Xu Z. H. (2004). Preparation of poly(1-vinylimidazole)-grafted magnetic nanoparticles and their application for removal of metal ions. *Chemistry of Materials*, **16**(10), 1977–1983.

Tavallali H. (2011). Alumina-coated magnetite nanoparticles for solid phase extraction of Cd in water samples. *ChemTech,* **3**, 1647–1651.

Thio B. J. R., Clark K. K. and Keller A. A. (2011). Magnetic pollen grains as sorbents for facile removal of organic pollutants in aqueous media. *Journal of Hazardous Materials*, **194**, 53–61.

Tural B. (2010). Separation and preconcentration of boron with a glucamine modified novel magnetic sorbent. *Clean–Soil, Air, Water*, **38**(4), 321–327.

Wang H., Keller A. A. and Clark K. K. (2011). Natural organic matter removal by adsorption onto magnetic permanently confined micelle arrays. *Journal of Hazardous Materials*, **194**, 156–161.

Wang J., Zheng S., Shao Y., Liu J., Xu Z. and Zhu D. (2010). Amino-functionalized Fe_3O_4@ SiO_2 core–shell magnetic nanomaterial as a novel adsorbent for aqueous heavy metals removal. *Journal of Colloid and Interface Science*, **349**(1), 293.

Wang P., Shi Q. H., Shi Y. F., Clark K. K., Stucky G. D. and Keller A. A. (2009). Magnetic permanently confined micelle arrays for treating hydrophobic organic compound contamination. *Journal of the American Chemical Society*, **131**(1), 182–188.

Wang Y., Liu J., Wang P., Werth C. J. and Strathmann T. J. (2014). Palladium nanoparticles encapsulated in core-shell silica: a structured hydrogenation catalyst with enhanced activity for reduction of oxyanion water pollutants. *ACS Catalysis*, **4**(10), 3551–3559.

Warner C. L., Addleman R. S., Cinson A. D., Droubay T. C., Engelhard M. H., Nash M. A., Yantasee W. and Warner M. G. (2010). High-performance, superparamagnetic, nanoparticle-based heavy metal sorbents for removal of contaminants from natural waters. *Chemsuschem*, **3**(6), 749–757.

Xie L. J., Jiang R. F., Zhu F., Liu H. and Ouyang G. F. (2014). Application of functionalized magnetic nanoparticles in sample preparation. *Analytical and Bioanalytical Chemistry*, **406**(2), 377–399.

Yamaura M., Camilo R. L., Sampaio L. C., Macedo M. A., Nakamura M. and Toma H. E. (2004). Preparation and characterization of (3-aminopropyl) triethoxysilane-coated magnetite nanoparticles. *Journal of Magnetism and Magnetic Materials*, **279**(2–3), 210–217.

Yang F., Shen R., Long Y. M., Sun X. Y., Tang F., Cai Q. Y. and Yao S. Z. (2011). Magnetic microsphere confined ionic liquid as a novel sorbent for the determination of chlorophenols in environmental water samples by liquid chromatography. *J Environ Monitor*, **13**(2), 440–445.

Yantasee W., Warner C. L., Sangvanich T., Addleman R. S., Carter T. G., Wiacek R. J., Fryxell G. E., Timchalk C. and Warner M. G. (2007). Removal of heavy metals from aqueous systems with thiol functionalized superparamagnetic nanoparticles. *Environmental Science and Technology*, **41**(14), 5114–5119.

Zargar B., Parham H. and Hatamie A. (2009). Fast removal and recovery of amaranth by modified iron oxide magnetic nanoparticles. *Chemosphere*, **76**(4), 554–557.

Zhai Y. M., Zhai J. F., Zhou M. and Dong S. J. (2009). Ordered magnetic core-manganese oxide shell nanostructures and their application in water treatment. *Journal of Materials Chemistry*, **19**(38), 7030–7035.

Zhao X. L., Shi Y. L., Ca Y. Q. and Mou S. F. (2008a). Cetyltrimethylammonium bromide-coated magnetic nanoparticles for the preconcentration of phenolic compounds from environmental water samples. *Environmental Science and Technology*, **42**(4), 1201–1206.

Zhao X. L., Shi Y. L., Wang T., Cai Y. Q. and Jiang G. B. (2008b). Preparation of silica-magnetite nanoparticle mixed hemimicelle sorbents for extraction of several typical phenolic compounds from environmental water samples. *Journal of Chromatography A*, **1188**(2), 140–147.

Zhao X. L., Wang J. M., Wu F. C., Wang T., Cai Y. Q., Shi Y. L. and Jiang G. B. (2010). Removal of fluoride from aqueous media by Fe_3O_4@Al(OH)(3) magnetic nanoparticles. *Journal of Hazardous Materials*, **173**(1–3), 102–109.

Zhou Y.-T., Nie H.-L., Branford-White C., He Z.-Y. and Zhu L.-M. (2009). Removal of Cu^{2+} from aqueous solution by chitosan-coated magnetic nanoparticles modified with α-ketoglutaric acid. *Journal of Colloid and Interface Science*, **330**(1), 29–37.

Chapter 3

Rational design of functional nanoporous materials to confine water pollutant in controlled nano-space

Swasmi Purwajanti, Jie Yang, Xiaodan Huang, and Chengzhong Yu

3.1 INTRODUCTION

Inadequate access to clean water is one of the most pervasive problems afflicting people throughout the world. In 2013, World Health Organisation (WHO) reported that there were more than 0.78 billion people around the world who did not have access to clean and safe water resources. It is estimated that within couple of decades the current water supply also will decrease by one-third (Amin *et al.* 2014). The problem with water is expected to grow worse in the coming decades and will be occurring globally even in regions which are currently considered as water-rich region (Shannon *et al.* 2008). The shortage of clean and safe water is mainly due to the pollution of ground/surface water by algal blooms, detergents, fertilizer pesticides, chemicals, heavy metals and so forth (Falconer, 2005; Fawell & Nieuwenhuijsen, 2003; Len Ritter & Patricia Keen, 2002; Rodriguez-Mozaz *et al.* 2004). Meanwhile the decrease of fresh water supply is also resulted from the exploitation of water resources for domestic, industry and irrigation purposes in many parts of the world (Shannon *et al.* 2008).

Pollutants found in water are generally classified into two categories, i.e. microbial pollutants and chemical pollutants and they may occur naturally or be derived from anthropogenic activity. The contamination of water by microbial pollutants can cause waterborne disease such as diarrheal disease, cholera and intestinal parasitic infection, which has been reported to become a leading cause of malnutrition (Lima *et al.* 2000). Meanwhile, chemicals pollutants may enter water through leaching, accidental spills, runoff, and atmospheric deposition and associate with a broad array of adverse health effects, including cancer, cardiovascular disease, neurological disease and miscarriage (Barrett, 2014; Calderon, 2000).

3.2 ARSENIC AND PHOSPHATE AS POLLUTANTS

Taking arsenic and phosphate as example chemical pollutants in water, their contamination to water has effected the environment and health of million people worldwide. Arsenic is one of the most toxic elements in the world and ubiquitously present in air, soil, natural water, mineral deposits and rocks in varying concentrations (Abernathy *et al.* 1999; Singh *et al.* 2015). The exposure of arsenic into environment has been reported worldwide and it has become a serious health threat to the human beings in some parts of the world (Argos *et al.* 2010). The arsenic crisis happened in Bangladesh has been described as "the largest mass poisoning of a population in history" by WHO (Argos *et al.* 2010). The release of arsenic from arsenic-rich minerals by geochemical factors is the main source of arsenic presence in the environment. Other sources are from anthropogenic activities i.e. through use of insecticides, herbicides, and phosphate fertilizers, semiconductor industries, mining and smelting, industrial processes, coal combustion, timber preservatives etc. (Mondal *et al.* 2006). Two main pathways for arsenic to enter in human beings are through the consumption of arsenic contaminated drinking water and from the intake of food-containing arsenic. Therefore, arsenic accumulation in food i.e. vegetables followed by ingestion may increase the level of daily human intake of inorganic arsenic (Fontcuberta *et al.* 2011). High level exposure of inorganic arsenic causes melanosis, leuco-melanosis, keratosis, hyperkeratosis, dorsum, non-petting edema, gangrene and skin cancer (Singh *et al.* 2015). Excessive and long term exposure above 0.05 mg L^{-1} of concentration may result in the development of arsenicosis which is defined as common term used for arsenic related health effects including skin problems, skin cancers, internal cancers (bladder, kidney, lung), disease of the blood vessel of the legs and feet, and possibly diabetes (WHO, 2011). Considering the effect on human health, WHO set guidelines that the recommended limit of arsenic in drinking water is 0.01 mg L^{-1}. By applying this provisional guideline, it is estimated that the total population being at risk of arsenic exposure is 145 million (Ravenscroft *et al.* 2009).

Meanwhile phosphorus, which exists as pentavalent form in natural waters, can be found in soil, water and sediments and are widely used in the areas of agriculture and industry (Yu & Chen, 2015). The main source of phosphorus in biosphere is from weathering or the erosion of phosphorus-containing minerals such as calcium phosphate and apatite. Another major source of phosphorus entering water includes drainage from agricultural land, excreta from livestock, municipal and industrial effluent and urban drainage (Yeoman *et al.* 1988). However, phosphorus is the limiting nutrient in most freshwater lakes, reservoirs and rivers. The phosphorus enrichment in water bodies through excessive discharge of phosphate-containing wastewater and natural processes can lead into detrimental eutrophication process and subsequently degrade the water quality. Eutrophication can affect the ecological balance of water, as there will be an abundance of algal "blooms" which can deplete dissolved oxygen in water that further can cause the death of fish. Therefore, WHO regulates the maximum discharge limit for phosphorus of 0.5–1 mg L^{-1} and also USEPA recommend that the maximum level does not exceed 50 μg L^{-1} in order to prevent eutrophication problem (Yu & Chen, 2015).

3.3 CURRENT DEVELOPED TECHNIQUES FOR ARSENIC AND PHOSPHATE REMOVAL

In order to alleviate the adverse effect from arsenic and phosphate, there have been continuous demands to develop technologies to reduce both the toxicity and the concentration of contaminants either in contaminated ground/surface waters or in municipal and industrial discharges. For arsenic, current developed techniques for remediation of arsenic contaminated water include oxidation/precipitation technique, coagulation/co-precipitation technique, sorption/ion exchange, membrane technology and other techniques such as solvent extraction, foam flotation and bioremediation (Mohan & Pittman, 2007). Mohan and Pittman (2007) on their critical review on arsenic removal from water/wastewater using adsorbents has discussed advantages and disadvantages of the above removal methods. Oxidation/precipitation method is selected due to the chemical property of arsenic. In natural waters, arsenic normally occurs in the oxidation states, i.e. +III (arsenite) and +V (arsenate). The removal of As(III) is more difficult than the removal of As(V). Therefore, As(III) has to be oxidized to As(V) prior to its removal (Bissen & Frimmel, 2003; Xu *et al.* 2008). In oxidation/precipitation method, air and some chemicals can be used as oxidizing agents, followed by precipitation using metal salts (Borho & Wilderer, 1996). This method is simple and relatively cost-efficient – depending on the oxidizing agent used. It can provide *in situ* arsenic removal and also can oxidize other inorganic and organic constituents in water simultaneously. However, oxidation/precipitation technique mainly removes only As(V) if incomplete oxidation occurs. Thus, it needs efficient control of the pH and oxidation steps. Besides, this method is also limited by the formation of harmful by-products in oxidation steps (Bissen & Frimmel, 2003; Mohan & Pittman, 2007).

Coagulation/co-precipitation technique for removal of arsenic is usually performed by using metal coagulants such as alum ($Al_2(SO_4)_3$), ferric chloride ($FeCl_3$) and lime (Jiang, 2001; Mohan & Pittman, 2007). This method is simple in operation and requires relatively low capital cost and also applicable for a wider range of pH. However, toxic sludge produced during coagulation is unavoidable and provides low removal capacity of arsenic since it is only more effective for removal As(V) than As(III). In some cases, to enhance the effectiveness, pre-oxidation may be required prior to coagulation process. Other developed techniques for arsenic removal are membrane techniques which includes nanofiltration, reverse osmosis and electrodialysis (Ning, 2002). It is reported that membrane technologies are effective in decreasing arsenic concentration to the target value of 10 μg L^{-1} with oxidation pre-treatment (Elcik *et al.* 2013). Nevertheless to apply this technology, it needs very-high capital and running cost as well as maintenance and the toxic wastewater may be produced from eletrodialysis. In addition to that this technology has not been shown to be feasible for developing world (Jiang, 2001).

At the same time, technologies for phosphorus removal have also been developed, providing opportunities for phosphorus recycling and sustainability. The development started at 1950s, as a response to the issue of eutrophication and the need to reduce the level of phosphorus entering waterbodies (Morse *et al.* 1998). Initially developed technique is chemical precipitation and it remains the leading technology today. Chemical precipitation is carried out by adding of divalent or trivalent metal salts to wastewater to generate insoluble metal phosphate precipitations. The precipitations could be later settled out by sedimentation. The drawback of this technique is that it fails in providing a more valuable and consistent sludge for recycling phosphorus to agriculture and industry. Moreover, the sludge requires post-treatment before it can be placed into landfill. Biological treatment technology is also developed and widely used in the treatment of phosphate-contaminated water. This technique has advantages such as relatively low cost and high efficiency. However, nutrients such as iron must be added in the water in order that biological reaction can occur. Meanwhile, ion exchange and reverse osmosis can be applied only when phosphate concentration is low and no other contaminant exists, which limits their application in large wastewater treatment plants (Yu & Chen, 2015). The application of membrane technology on phosphorus removal has shown good performance to remove not only phosphorus in total suspended solids but also the dissolve phosphorus. According to Reardon's report, membranes technology can reduce the total phosphorus concentration to less than 1 mg L^{-1} in their effluent (Reardon, 2006). However, this technology has several significant limitations for phosphorus removal including quite high initial investments, expensive membrane replacement, high energy consumption and large amounts of sludge produce (Jiang *et al.* 2004).

3.4 ADSORPTION AS AN ALTERNATIVE APPROACH FOR ARSENIC AND PHOSPHATE REMOVAL

For both arsenic and phosphate pollutants removal, adsorption method offers several advantages over the other developed and commercialized technologies. Adsorption is an easy and flexible method with high removal efficiency. To be specific, advantages of adsorption technology are: (i) it usually does not require large volume of additional chemicals; (ii) it is easy to be applied in point of entry or point of use pollutants removal processes (Jang *et al.* 2008); (iii) it produces no harmful by product and may provide sludge-free operations (Genç-Fuhrman, 2004; Zhang *et al.* 2005). For phosphate especially, adsorption technology does not have the requirement for carbon/nitrogen or other nutrients.

Currently, a large number of commercial adsorbents for arsenic removal are available in the market. The class of commercially available adsorbents ranges from biological materials, mineral oxides, different soils, silica materials, activated carbons to polymer resins. Here we list some examples of commercialized materials that have superior performances compared to the others. For silica class material, amended silicate adsorbent developed by ADA Technologies was able to achieve removal capacities of 2 mg g^{-1} at initial concentration of 50 μg L^{-1} and about 40 mg g^{-1} if the initial concentration increased to 1000 μg L^{-1}. In addition to that, this type of adsorbent can reduce the arsenic concentration as high as 1000 μg L^{-1} to 10 μg L^{-1} within 30 min (Frazer, 2005). ADA technologies also developed and commercialized arsenic point of use and point of entry removal systems using the same type of adsorbent. Adsorbsia® GTO®, developed by The Dow Chemical Company, is granular titanium oxide with strong affinity for arsenic. At 50 μg L^{-1} and pH 7, this media has shown improved capacity towards As(III) and As(V) with capacity of 3–4 mg g^{-1} and 12–15 mg g^{-1}, respectively. Introduced by Engelhard, ARM 200 is an efficient and cost-effective arsenic removal adsorbent, specially designed for use in household, industrial and water utility filtration systems. Using NSF 53 protocol testing at pH 8.5, ARM 200 can treat up to 2000 gallons and still can generate high quality effluent that is under maximum concentration limit of 10 μg L^{-1}. Another commercialized media is ferric/lanthanum hydroxide compound deposited onto a diatomaceous earth substance to provide high surface area. It was developed by EaglePicher Filtrationandminerals Inc. This media was proposed to remove arsenite and arsenate without a required chemical pre-treatment. Since within this media arsenic forms permanent bonds, the removal is irreversible.

Meanwhile for phosphorus removal, not many commercialized adsorbent have been developed to remove phosphate from contaminated waters. In 2002, a screening of several commercial adsorbents for phosphorus removal in fresh and seawater aquarium was conducted by Jekel, M. G. an co-workers (Jekel *et al.* 2002). Three commercial adsorbents tested ranged from iron-based (RowaPhos) to aluminium-based (ElimiPhos) materials, in forms of zeolite (AntiPhos, PhosEx) or ceramics (Phosphate Sponge). Iron-based material, in the form of granulated ferric hydroxide (GFH), showed a high affinities for phosphate adsorption compared to other commercial materials (Jekel *et al.* 2002). Maximum adsorption capacity that can be reached by GFH was 17.3 mg g^{-1} when treating tap water and 11.3 mg g^{-1} when treating sea water.

3.5 NANOPOROUS MATERIAL AS PROMISING ADSORBENT

Although there are a wide range of commercially available adsorbents for arsenic and phosphorus removal, their adsorption capacities towards those pollutants are still relatively low. Therefore, efficient and inexpensive arsenic and phosphorus adsorbents are still highly desirable. Nanoporous materials have been utilized in many applications, including catalysis, drug delivery, imaging, purification and separation. They also have been considered as potential adsorbents, due to their high intrinsic surface area, high pore volume, adjustable framework, as well as the unique surface properties (Wu & Zhao, 2011). These characteristics, especially the high surface area, can fulfil the requirements for a good adsorbent. According to IUPAC, porous material can be classified into three categories, i.e. microporous (pore diameter <2 nm), mesoporous material (pore diameter within 2–50 nm) and macroporous materials (pore diameter >50 nm) (Davis, 2002; Zhang *et al.* 2009). Examples of nanoporous materials that have utilized as adsorbents in water treatment are classified into two major groups i.e. inorganic materials and carbon based materials (Kanel & Nepal, 2009). Inorganic material such as mesoporus silica has been considered to be a good adsorbent for adsorbates such as metal ions and amino acids (Burke *et al.* 2009; O'Connor *et al.* 2006; Sawicki & Mercier, 2006; Wu & Zhao, 2011). This advantage is attributed to its high specific surface area, controlled pore diameters and controlled morphology, e.g. spheres, rods and disk (Hosni *et al.* 2007). In addition to that, mesoporus silica is also considered as mechanically robust, non-toxic and environmentally friendly. Meanwhile carbon based materials and its new generations have been extensively used for liquid phase adsorption, separation and purification processes due to relatively large surface area and exhibiting novel electronic and chemical property which can enhance the adsorption processes (Gracia-Espino *et al.* 2012; Kanel & Nepal, 2009).

Furthermore, in environmental research, particularly in the field of water treatment, nanoporous materials have a strong impact owing to their controllable pore size and high surface area especially when they are used in nano-size. The adsorption capacity is one of the most important criteria for a good adsorbent, which is directly related to specific surface area and pore size. Thus, the precise control of pore size as well as porous structure is of great importance. In this chapter, we will summarise the development of functional nanoporous materials applied in arsenic and phosphorus adsorptions. Both materials design and adsorption investigation will be introduced. The correlations between structural feature and adsorption performance will be discussed to summarize current findings and new understandings in this research area.

3.6 FUNCTIONAL NANOPOROUS MATERIAL FOR ARSENIC REMOVAL

Currently, a large number of different adsorbents capable of removing arsenic have been reported. Among them, iron oxide-based adsorbents are the most promising ones owing to their high adsorption capacity and excellent selectivity (Chen *et al.* 2007; Katsoyiannis & Zouboulis, 2002; Wu *et al.* 2012). Iron oxide nanoparticles (maghemite and magnetite) with size 12 and 3.7 nm showed significantly increased arsenic capture performance compare to large particle size iron oxide. The improved performances may result from more adsorption sites being exposed to arsenic species (Tuutijärvi *et al.* 2009; Yavuz *et al.* 2006b; Yean *et al.* 2005). Nevertheless, nanoparticles tend to aggregate into large agglomerates, easily resulting to decreased adsorption performance. Nanoparticles would also cause the difficulties in separation and/or diffusion if applied in the wastewater treatment. In order to overcome these problems, it is desirable to use nanostructured materials as the matrix to disperse iron oxide nanoparticles, in order to take advantages of small particle size and high surface area, which is important for adsorption applications (Baikousi *et al.* 2012; Lu *et al.* 2010; Tuček *et al.* 2014).

Graphene and graphene based nanostructures have received great attention recently as proven candidates to act as supports for nanoparticles immobilization for developing advanced adsorbents in environmental applications (Andjelkovic *et al.* 2015). At the first attempt, water-dispersible magnetite-reduced graphene oxide composite is synthesized by Chandra and co-workers in 2010 for arsenic removal (Chandra *et al.* 2010). This hybrid composite was synthesized via a chemical reaction between graphene oxide and iron precursors with ammonia and hydrazine at pH 10 and 90°C. The presence of magnetite in this reduced graphene oxide composite was confirmed by X-ray photoelectron spectroscopy (XPS) and wide-angle X-ray diffraction (XRD). The magnetite particle size decorated the reduced graphene oxide is averagely 10 nm. This composite possessed superparamagnetic property at room temperature and could be separated by an external magnetic field. The adsorption capacity of magnetite-reduced graphene oxide for As(III) and As(V) was investigated in batch adsorption process using the initial concentration of 50 mg L^{-1}. The adsorption capacity obtained from the batch adsorption was 13.10 and 5.83 mg g^{-1} for As(III) and As(V), respectively. The adsorption data fitted both Langmuir and Freundlich isotherm models. The interesting finding is that the removal capacity of As(III) is higher compared to that of As(V). Similar observation was also found in ferric-oxide loaded polymeric adsorbents (Zhang *et al.* 2008). In addition to that, the pseudo-second order kinetic model can describe the adsorption behaviour for both As(III) and As(V). The high adsorption capacity for As(III) and As(V) could be attributed to the increased adsorption sites in non-aggregated bare magnetite nanoparticles decorated on the reduced graphene oxide.

Developing three-dimensional nanostructures is one of the most facile strategies to obtain novel materials with high surface-to-volume ratio arising from unique morphology owning advantages for several applications, especially for water purification (Georgakilas *et al.* 2012; Kim *et al.* 2009). Recently, three-dimensional (3D) nanostructures based on graphene (G) as matrix decorated with one-dimensional carbon nanotubes (CNT) and iron oxide nanoparticles (G-CNT-Fe) for effective arsenic removal from the contaminated water were developed by Vadahanambi and co-workers (Vadahanambi *et al.* 2013). One-pot single step microwave method, involving high purity expandable graphite, ferrocene and azodicarbamide as reagents, was employed for the synthesis enabling the mass production of 3D G-CNT-Fe. During the synthesis in microwave environment, three different phenomena occurred simultaneously i.e. expansion of graphene oxide to graphene nano-worms, formation of iron oxide nanoparticles from ferrocene decomposition and vertical growth of CNT no graphene surface. TEM result revealed that the iron oxide particles are well distributed, attached on the surface of both graphene and CNT. The results from XPS indicated that the particles existed in a mixed state of Fe_2O_3 and FeOOH. The well-distribution of iron oxide nanoparticles is very beneficial in the removal of heavy metals from drinking water such as arsenic. The functionality of G-CNT-Fe towards arsenic removal from contaminated water was proved from XPS analysis. A survey scan of core level binding energy of arsenic adsorbed G-CNT-Fe showed that there was a considerable wakening of Fe_{3s} peak suggesting the anionic adsorption of arsenic and the formation of arsenate complexes on the surface. More detailed electronic structures of these arsenic complexes were revealed in high resolution As_{3d} deconvoluted spectra which showed three prominent peaks at 43.6, 44.5, and 45.4 eV associated with As—O bonds, As_2O_3, and As_2O_5, respectively. When G-CNT-Fe was compared with 2D-iron-ecorated graphene hybrids (Sridhar & Oh, 2011), the capacity of arsenic removal of G-CNT-Fe was almost double at all concentrations of arsenic. The authors attributed the outstanding arsenic removal performances of G-CNT-Fe to its novel three-dimensional nanostructure. They claimed that high meso-porosity and open porous networks of G-CNT-Fe resulted in faster molecular diffusion and enhanced the accessibility of iron oxide nanoparticles. However, this conclusion could be a bit contentious, because there wasn't direct evidence to prove the mesoporous structure. The porosity analysis was almost missing and the comparison study was not sufficient.

A good example of iron oxide incorporated nanoporous arsenic adsorbents was reported by Wu and co-workers in 2012. They reported a facile, controllable and versatile approach for the synthesis of ordered mesoporous carbon encapsulating iron oxide nanoparticle with a high concentration and homogeneous distribution for capturing arsenic (Wu *et al.* 2012). The synthesis employed an ammonia-atmosphere pre-hydrolysis post-synthetic route and can achieve very high iron oxide content up to 52 wt%. Referring to the schematic illustration presented in Figure 3.1, the preparation of this composite started from preparation of the ordered bimodal mesoporus carbon matrix (Figure 3.1a) which was synthesized according to the procedure reported by Liu and co-workers (Liu *et al.* 2006). The ordered bimodal mesoporus carbon matrix was functionalised by a surface oxidation with acidic ammonia persulfate solution at 70°C for 12 h. The obtained functionalized carbon matrix was then impregnated with 20% ethanolic solution containing iron salt until the solvents was evaporated (Figure 3.1b). Next step was the in situ hydrolysis in 14 wt% ammonia solution for 3 h at 60°C, continued with pyrolysis at 300–600°C for 1 h to obtain the final mesoporous Fe_2O_3@C encapsulates (Figure 3.1c and e).

High-magnification transmission electron microscopy image could reveal that iron oxide nanoparticles were spatially and selectively located in the primary mesopores of the carbon matrix without formation of any aggregation. The high-resolution elemental mapping confirmed that the Fe, C and O elements were quite faithfully correlated with the ordered mesostructured domain, without any formation of large particle outside the mesopores, indicating that the nanoparticles were exclusively deposited inside the mesopore system. The result from N_2 sorption isotherm also confirmed that almost all nanoparticles were selectively deposited in the primary mesopores. The results showed that the pore volume for the primary mesopores decreases dramatically to only one-fifth of that in the pristine carbon matrix, while the interconnected pore remained constant even at high content of incorporated iron oxide. In addition to that, there was also a slight decrease of ~0.8 nm for primary mesopores within carbon matrix. It also implied that iron oxide nanoparticles were selectively deposited in the primary mesopores. The key concept for this material design is selectively load nanoparticles with a very high concentration into primary mesopore (5.6 nm) of surfactant-templated ordered bimodal mesoporous carbon matrix, while leaving its inter-connected mesopores (2.3 nm) empty to retain an open pore network. Thus, fast molecular diffusion/ transportation can be maintained along with high content of active materials. This material showed high capacities of arsenic removal because of the synergetic effects of high and open-pore mesoporosity and uniformly dispersed high content of iron oxides. The mesoporous iron oxide encapsulated in carbon matrix also possessed magnetic property that would advantage the separation process after adsorption by simply applying external magnetic field.

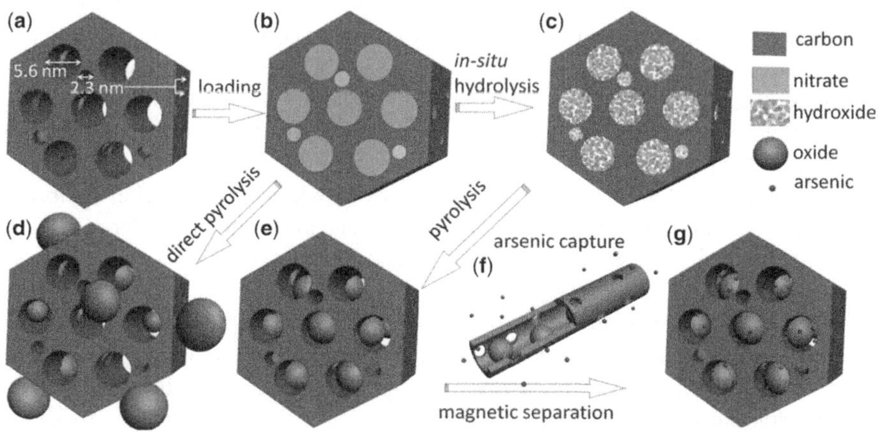

Figure 3.1 Schematic illustration of the synthesis and arsenic adsorption processes for ordered mesoporous Fe_2O_3 (a) the bimodal mesoporous carbon, (b) carbon loaded with hydrated iron nitrate precursor, (c) carbon loaded with iron hydroxide obtained by in situ hydrolysis under ammonia atmosphere, (d) iron oxide@carbon composites obtained by direct pyrolysis, (e) the Fe_2O_3@C encapsulates obtained by pyrolysis following the pre-hydrolysis, (f) arsenic capture, and (g) arsenic-enriched encapsulates. (Wu *et al.* 2012) Copyright 2012 Wiley-VCH.

When this material was applied for arsenic removal, it exhibited impressive capacities of ~27.7 mg g^{-1} for sample pyrolisized at 300°C which were overwhelmingly higher than that of carbon matrix with ~1.4 mg g^{-1}. The significant increase in the As(III) adsorption capacity can be ascribed to the formation of stable mono or/a bidentate complexes on the surface of iron oxides in addition to physisorption by the carbon matrix. Compared to other reported arsenic adsorbents, it was also having quite competitively larger adsorption capacity (Chandra *et al.* 2010; Chen *et al.* 2007; Gu *et al.* 2007; Gu *et al.* 2005; Zhong *et al.* 2006). For As(V) adsorption on the carbon matrix and Fe_2O_3@C encapsulates, similar results was found. In adsorption isotherm study, As(III) and As(V) adsorption behaviours on the mesoporous encapsulates followed the Langmuir model, indicating a monolayer adsorption taking place with very high adsorption capacity and excellent saturation capacities. The maximum capacities reached by Fe_2O_3@C encapsulates were 29.4 mg g^{-1} and 17.9 mg g^{-1} for As(III) and As(V), respectively. The different adsorption capacities towards As(III) and As(V) could be probably attributed to their different interactions with iron oxide. From kinetic study, this novel adsorbent showed a high removal efficiency. It could remove 90% of 5 ppm As(III) in aqueous solution within 10 min even using relatively low dosage adsorbent of 1.0 mg mL^{-1}. The remarkable arsenic capture performance was resulted from the combination of several unique features owned by this novel material. High mesoporosity and open pore network of the carbon matrix can assist fast molecular diffusion and enhance the accessibility of iron oxides. At the same time, discrete active sites for arsenic adsorption and the magnetic property for better separation and regeneration can be provided by the high content, uniformly dispersed and spatially separated iron oxide nanoparticles confined in the pre-defined mesopores.

Encapsulation of high content magnetic iron oxide into sepiolite, a fibrous magnesium silicate clay mineral with interpenetrated and regular mesopore system, for the adsorption of arsenic pollutant was also reported by Tian and co-workers (Tian *et al.* 2015). This reported novel work tried to address the challenge of how to disperse the encapsulated iron oxide within the porous matrix in controlled manner. The preparation involved a simply pre-carbonization treatment of sepiolite, followed by a facile high-temperature decomposition of iron precursors onto pre-treated sepiolite in liquid polyols, in which *in situ* anchorage of mono-dispersed iron oxide nanocrystals would take place. Iron oxide nanocrystals were selectively deposited in sepiolite nanofiber surface and showed narrow size distribution from 5 to 15 nm with an average size of ~9 nm. Elemental mapping of the composite further elucidated the uniform distribution of iron oxide nanocrystals along the sepiolite nanofibers at the micron scale. N_2 adsorption-desorption result showed that iron oxide nanocrystals coverage on the sepiolite surface reduced the specific surface area from 298 to 105 m^2 g^{-1}. Detailed analysis on N_2 sorption result revealed that the surface area decrease was mainly due to the micropores blockage, implying abundant mesoporous structure remained. This composite exhibited a strong superparamagnetic property at room temperature, which is preferable for many practical applications especially for separation after adsorption.

Investigation in the adsorption property of As(III) of encapsulated magnetic iron oxide into sepiolite was conducted. The data of As(III) adsorption fitted Langmuir model better than Freundlich model and gained the maximum adsorption capacities of As(III) of 35.15 mg g^{-1} and 50.35 mg g^{-1} for composite calcined at 250 and 500°C. These capacity values were much higher compared to those in other reported works (Chandra *et al.* 2010; Ma *et al.* 2013; Vadahanambi *et al.* 2013; Wu *et al.* 2012; Yu *et al.* 2013). When it was applied to remove arsenic concentration as much as 1 mg L^{-1} at 298 K, the adsorption occurred rapidly during initial 1 hour and achieved the equilibrium at 3 hour. This rapid adsorption ability was resulted from large number of available adsorption sites. The adsorption taken place on the surface of the encapsulated magnetic iron oxide was described as pseudo-second order kinetic model indicating the as-obtained composite had a large adsorption capacity for As(III) and possessed the potency to be used as an excellent adsorbent for As(III) in groundwater treatment. In the presence of competing anions such as PO_4^{3-}, HCO_3^-, NO_3^-, SO_4^{2-}, CO_3^{2-}, F^- and Cl^-, there was no diverse influence on the As(III) removal excepting fluoride and phosphate. For the removal of low concentration of As(III), it also showed remarkable performance. The adsorbent can reduce the As(III) concentration to lower than 5 μg L^{-1} that meet the minimum concentration level of arsenic in ground water. The real groundwater from Jianghan Plain containing 456 μg L^{-1} of arsenic was also treated with magnetic iron oxide encapsulated sepiolite. The result showed that this composite could exhibit high removal ratio of 96.4% to meet the safe As concentration level. The high arsenic adsorption performance of magnetic iron oxide encapsulated sepiolite nanostructures could be owing to its fascinating features, such as high porosity and open pore network of the sepiolite which can promote fast arsenic diffusion and transportation and improve the accessibility of iron oxide. Additionally, the presence of decorated iron oxide nanocrystals, which was high-content, uniformly-dispersed and spatially separated, would increase the availability of active sites for capturing arsenic.

The research demonstrating the confinement of arsenic in controlled nanopores was performed by Yang and co-workers (Jie Yang *et al.* 2014). They prepared a novel composite of γ-Fe_2O_3 encapsulated Macroporous siliceous foam (MOSF) as the arsenic adsorbent. This preparation involved the impregnation of MOSF with $Fe(NO_3)_3.9H_2O$ by solvent evaporation. A solvothermal reaction was conducted in an autoclave containing acetic acid (≈20 ml) at 80°C for 3 h. The final γ-Fe_2O_3@MOSF was obtained after the pyrolysis of the solvothermal product at 400°C for 30 min under a 5% H_2/95% Ar atmosphere and the subsequent calcination at 150°C for 2 h in air. Three different weight percentages of γ-Fe_2O_3 were incorporated into MOSF matrix i.e. 11.8, 21.0, and 34.8% respectively. The observation on morphologies of MOSF before and after encapsulation of iron oxide nanoparticles showed that the macroporous foam-like structure of MOSF was well maintained. Observations through TEM were presented in Figure 3.2a–d, the distribution of Fe_2O_3 nanoparticles can be clearly seen through the whole sample (Figure 3.2b–c).

The size distribution histogram of nanoparticles showed a broad distribution of particles from 2–13 nm with an average size of ~6 nm. The dark-field scanning TEM technique further confirmed the successful γ-Fe_2O_3 loading. Many bright dots can be found in the dark domain of MOSF containing Si and O, indicating the existence of heavy atoms, namely Fe (Figure 3.2d). Later, a fast developing technique for the 3D imaging of complex material was applied to check whether nanoparticles was spatially well-dispersed and attached on the silica walls or just existed as aggregates, which cannot be confirmed using traditional 2D imaging technique such as TEM. The result demonstrated that only separated and well-dispersed dark dots could be found from xy, xz and yz direction, revealing γ-Fe_2O_3 nanoparticles mainly attached on the silica wall surface without aggregation. Non-aggregated small γ-Fe_2O_3 nanoparticles anchored on the pore wall was resulted from the homogenous distribution of iron precursor inside the pores of MOSF. This can be explained from the synthesis process, in which trinuclear Fe(III)-acetate complexes were formed and fixed on the wall surface of MOSF after impregnation and exposure to vapors of acetic acid. Further calcination at 400°C resulted in the thermal decomposition of Fe(III)-acetate complexes and the formation of nuclei γ-Fe_2O_3. Since the amount of available iron precursors in a local area was limited, the growth of γ-Fe_2O_3 nanoparticles was restricted. Hence, only non-aggregated small γ-Fe_2O_3 nanoparticles could be found anchoring on the silica wall surface of MOSF.

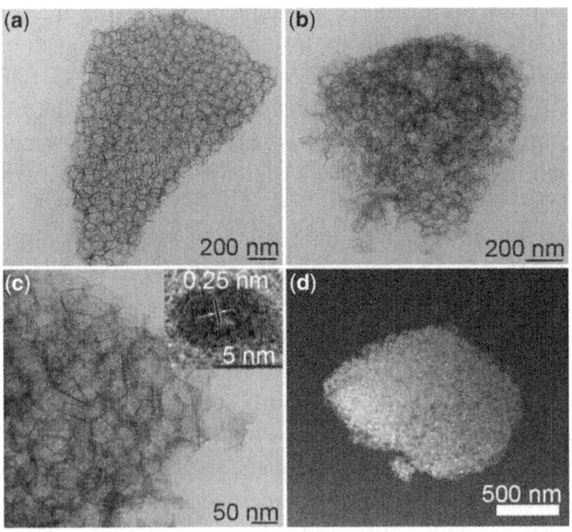

Figure 3.2 TEM images of (a) pristine MOSF, (b) γ-Fe_2O_3@MOSF (34.8%) at a low magnification and (c) a higher magnification. (d) DF-STEM image of γ-Fe_2O_3@ MOSF (34.8%). Inset of (c) is the HRTEM image of γ-Fe_2O_3 nanoparticles inside the pores of MOSF. (Jie Yang *et al.* 2014) Copyright 2014 Wiley-VCH.

In addition to that, the presence of open macroporous structure even at high γ-Fe_2O_3 content of 34.8 wt% can still be detected by nitrogen sorption analysis. A noticeable reduction of surface area from 250 to 155 m^2 g^{-1} and pore volume from 1.57 to 0.49 cm^{-3} g^{-1} can be clearly observed with the increase of γ-Fe_2O_3 content from 11.8 to 34.8 wt%. The open macroporous structure together with a relatively high surface area and high pore volume would be beneficial for the fast diffusion and transportation of arsenic species and consequently enhance the adsorption kinetics. Evaluation on arsenic performance of γ-Fe_2O_3@MOSF was carried out using batch mode operation. When contaminated water model containing either As(III) or As(V) at concentration of 600 mg L^{-1} and 560 mg L^{-1} is used, eligible amount of arsenic could be adsorbed by pristine MOSF with adsorption capacities of 8 mg L^{-1} for As(V) and 16 mg L^{-1} for As(III). Meanwhile, γ-Fe_2O_3@MOSF can significantly improve the adsorption capacity towards As(III) and As(V), adsorption capacity that can be achieved ranging from 112 to 248 mg g^{-1} for As(V) and 144 to 320 mg g^{-1} for As(III). Both results of arsenic adsorptions were competitively larger than those of reported arsenic adsorbents (Cao *et al.* 2012; Mou *et al.* 2011; Tuutijärvi *et al.* 2009; Wu *et al.* 2012; Yavuz *et al.* 2006a; Zhang *et al.* 2013; Zhong *et al.* 2006). The adsorption isotherms for both As(III) and As(V) followed Freundlich adsorption isotherm, which is considered as a multilayer adsorption process. (Mou *et al.* 2011). Pseudo-second order kinetic model can best describe of arsenic adsorption rate towards either high or low concentration of As(III) or As(V), indicating that the adsorption process occurred through the chemical interaction (Ho & McKay, 2000). At high concentration, the equilibrium can be reached in the first 30 mins for both contaminants, suggesting the very fast As(III)/As(V) adsorption rate of γ-Fe_2O_3@MOSF. At low concentration of 100 μg L^{-1}, similar result was also found, demonstrating that both in high and low concentration range this adsorbent can rapidly remove the arsenic species from wastewater without pre-treatment. The authors also presented a negligible reduction of the adsorption capacity with the presence of competing anions such as Cl^-, CO_3^{2-}, SO_4^{2-} and PO_4^{3-}. The applicability of this composite for treating real wastewater containing arsenic 400 mg L^{-1} without pre-treatment was also tested and the results showed that this composite can still exhibit excellent performance with the adsorption capacities of 232 and 200 mg g^{-1} for As(III) and As(V), respectively. In brief, the high performance should be credited to the special features owned by γ-Fe_2O_3@MOSF composite such as the large open pores to enhance the diffusion and transportation of arsenic as well as the high-content and well dispersed iron oxide nanoparticle which provided sufficient active sites for adsorption to take place.

3.7 FUNCTIONAL NANOPOROUS MATERIAL FOR PHOSPHORUS REMOVAL

Surface modified mesoporous silica materials have been synthesized and used for phosphate removal. Functional modifications such as amino group functionalization and metal oxides doping have been shown to have an positive effect in increasing the adsorption capacity towards phosphate compared to unmodified silica (Ou *et al.* 2007; Saad *et al.* 2008; Shin *et al.* 2004; Zhang *et al.* 2010). Delaney *et al.* successfully removed phosphate from wastewater using metal oxides (Ti, Zr, Fe, and Al) doped ordered mesoporous silica (OMS) (Delaney *et al.* 2011). The preparation of metal oxide doped OMS employed a one-pot synthesis with the mixing of the triblock copolymer P123, tetramethoxysilane (TMOS) and metal salts source with the molar ratios of 20:1, 40:1, and 80:1. The mixture was aged at 313 K for one week. Then precipitates were separated, dried in oven and further calcined at 723 K in air for 5 h. The N_2 adsorption-desorption results showed that surface area and pore volume of the doped OMS were progressively reduced with the increase of dopant contents. When samples with 1 wt% dopant were utilized to adsorb phosphate in simulated wastewater with concentration 1 mg L^{-1}. The results revealed that Fe-OMS was the most promising adsorbent while un-doped OMS showed a negligible adsorption capacity towards phosphate. The 20:1 doped Fe-OMS was able to remove 94% after 10 hours adsorption, while the 40:1 and 80:1 sample can only remove 89% and 84% of the phosphate, respectively. This study also investigated the phosphate adsorption mechanism. Phosphates were found to be specifically adsorbed on reactive surface hydroxyl groups of inorganic material through a ligand exchange mechanism (Georgantas & Grigoropoulou, 2007). The adsorption largely depended on temperature and pH value. The increased temperature would be accompanied by a decrease in the phosphate uptake. That was consistent with a weak co-ordination mechanism, as the temperature increase could cause desorption of orthophosphate from the adsorbent surface. Meanwhile, the adsorption ability of metal oxide doped OMS reached to highest at neutral pH conditions. The complex solution chemistry of orthophosphate ion in either alkaline or acidic conditions would lead to largely decreased effectiveness.

Looking at the results above, metal doping is suggested as an effective way to improve the adsorption capacities of mesoporous silicas working as phosphate adsorbents. Research on this area has been growing exponentially (Huang *et al.* 2014). Compared with other metals, lanthanum (La) shows much more attractive features such as high adsorptive capability, high chemical stability, low toxicity and biocompatibility (Meiser *et al.* 2004). Zhang *et al.* reported the fabrication of a series La doped mesoporous MCM-41 by one-pot synthesis (Zhang *et al.* 2010). The doping molar ratios of Si/La were 100, 50 and 25. The adsorption capacities were found to be in accordance to the doping contents of doped La. Among these three samples, the Si/La molar ratio of 25 one demonstrated the highest Langmuir adsorption capacity of 21.1 mg g^{-1}.

Laanum-doped mesoporous hollow silica spheres were fabricated by Huang *et al.* and utilized as phosphate adsorbent for the first time (Huang *et al.* 2014). Mesoporous hollow silica spheres were synthesized using modified method reported by Vaudreuil *et al.* (Vaudreuil *et al.* 2001). Silica samples were incorporated with La using ethanol evaporation method at 60°C for 24 h. The composite samples further underwent a calcination to convert $La(NO_3)$ into lanthanum oxide. According to the La/Si molar ratio, the amount of La incorporated ranged from 1/50 to 1/2.5. With the increase of lanthanum incorporation amount, there was a decreasing trend in sample surface areas, average pore sizes and total pore volumes, which implied structural changes of the impregnated samples. This could be attributed to the impregnation of La inside the mesopore, which narrowed the pore sizes and reduced the pore volumes. It was also observed that after impregnation with La, all samples exhibited ordered mesopore in the shells with a similar thickness as that of un-impregnated hollow mesoporous silica spheres, indicating that all of the mesostructures can be retained after La impregnation. Investigation of the phosphate adsorption performance was conducted for all prepared samples at 25°C for 24 h. The results exhibited that hollow mesoporous silica doped with La/Si ratio of 1/5 (HMS-1/5) had the highest adsorption capacity of 47.89 mg g^{-1}. This could be ascribed to the suitable amount of La doped into the HMS-1/5 mesopores, providing sufficient active sites and good accessibility for phosphate adsorption. Compared with the other reported phosphate adsorbents, HMS-1/5 showed superior maximum adsorption capacity. This result was probably due to the great surface area of the HMS-1/5 adsorbent, i.e. 420 m^2 g^{-1}, together with its unique hierarchically macroporous-mesoporous structure that ensured the high accessibility of La active sites located in the mesoporous to phosphate anions. Evaluation on the La usage of HMS-1/5 showed a value of 0.91 indicating a high efficiency of La usage. Langmuir fitted the isotherm study of prepared samples, demonstrating adsorption processes were monolayer adsorptions. Some other works reported similar findings toward the characteristic of phosphate adsorption onto hollow mesoporous silicas (Biswas *et al.* 2007; Haghseresht *et al.* 2009; Li *et al.* 2009). The adsorption behaviour of phosphate onto the HMS-1/5 followed the pseudo-second order kinetic, which implies that the adsorption was likely governed by chemisorption. At low phosphorus concentration of approximately 2 mg L^{-1}, MHS-1/5 exhibited very fast adsorption. Within 14 min, 99.71% of the phosphate in the initial solution was removed and the final remaining phosphate in the initial solution reached 10.12 μg L^{-1}.

Although functionalized mesoporous materials had shown comparatively high phosphate adsorption capacities and fast adsorption kinetics, there was an apparent knowledge gap between structural features and phosphate adsorption properties. The confinement of phosphates in nanopores is of importance for the understanding of the fundamental correlation between pore structure and adsorption performance and for the rational design of phosphate adsorbents with superior performance. Recently, Yang and co-workers reported a series of detailed studies on the confinement of phosphates in nanoporous silica materials, which to a great extent bridged the gap on the structure–property relationship. In their first report (Jie Yang *et al.* 2011), a La functionalized mesoporous silica based material with high effective phosphate removal was developed. SBA-15 mesoporous silica was used as support matrix (Figure 3.3a). The design of this high effective adsorbent was based on several considerations i.e. (i) the careful choice of biocompatible La adsorbent as the functional component which can transform to $LaPO_4$ species with a very low K_{sp} (3.7×10^{-23}); (ii) fixing La species in the mesopores by controlling the calcination temperature to avoid the leaching and waste of functional component; and (iii) adsorbing $LaPO_4$ species in the mesopores by adjusting the pore size of mesoporous support to control the site where $LaPO_4$ nucleation and growth taken place. La precursor was incorporated into calcined SBA-15 using the ethanol evaporation method. The La loading content was controlled by varying the mass ratio between calcined SBA-15 and the impregnated lanthanum nitrate hexahydrates ($La(NO_3)_3{\cdot}6H_2O$) from 20 to 100%. The impregnated $La(NO_3)_3{\cdot}6H_2O$ was converted to Lanthanum oxide during the second calcination process to obtain the final products, La_xSBA-15, x denotes the mass ratio percentage of La. Adsorption study was carried out to evaluate the effect of La incorporated into SBA-15. The maximum adsorption amount (Q_{max}) of La_xSBA-15 was 0.529, 0.938, 1.307 and 1.362 mmolP g^{-1} (x = 20, 40, 80, and 100, respectively). It worth noting that increasing x from 80 to 100 resulted in only 4% increase in Q_{max}. From the molar ratio of P adsorbed versus La in each sample, it was also shown that La_{100}SBA-15 had the lowest P/La ratio, indicating that the high loading content of La in La_xSBA-15 materials might possibly lead to a decreased La usage efficiency. From the kinetic and isotherm study of the adsorption, it was suggested that the adsorption characteristic of phosphorus on La_xSBA-15 was homogeneously occurring through chemical interactions.

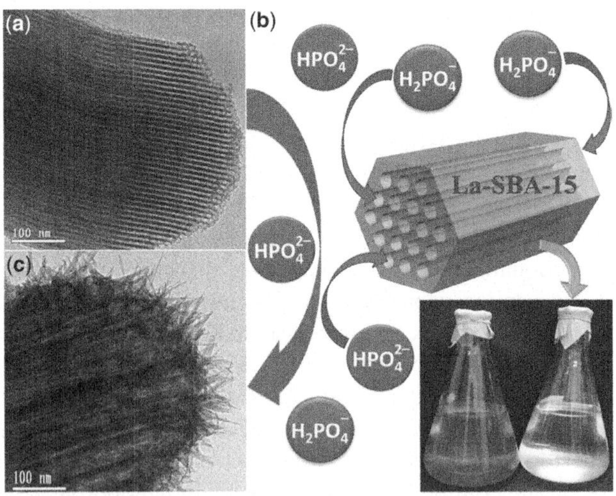

Figure 3.3 Schematic illustration of the P adsorption mechanism. TEM image (a) shows the typical structure La-SBA-15. (b) The adsorption of P and further nucleation of $LaPO_4$ species occurred in the mesopores, (c) the heterogeneous nucleation and further crystal growth lead to polycrystalline $LaPO_4$. Insert of (b) shows blue green algae cultivation without (left) and with (right) the addition of La-SBA-15. (Jie Yang *et al.* 2011) Copyright 2011 The Royal Society of Chemistry.

To further understand the P adsorption mechanism influenced by the confining nanospace (Figure 3.3b), investigations on phosphorus adsorption capacities of La_{100}SBA-15 with different pore sizes were carried out. La_{100}SBA-15 samples with controlled pore size were obtained by impregnation of $La(NO_3)_3 \cdot 6H_2O$ in SBA-15 supports synthesized with different hydrothermal temperatures. With the increase of pore sizes from 5.35 to 8.87 nm, the P adsorption capacities increased from 0.497 to 1.472 $mmol_P\ g^{-1}$. TEM was employed to monitor the evaluation of $LaPO_4$ species in La_{100}SBA-15 as a function of contact time (t). It was shown that at 370 min when 95% P was adsorbed, a few needle-like nano-rods can be observed growing out from the pores channels (Figure 3.3c). After 24 hr, large amounts of such nano-rods can be found. The diameters of such nano-rods were ~5 to 8 nm, close to the diameter of channel-like pore in La_{100}SBA-15. From HRTEM, it was confirmed that the needle-like structures were $LaPO_4$ nanorods. For comparison, smaller pore size mesoporous silica material, MCM-41, was used as the support. It showed that at 24 hr, $LaPO_4$ species formed mainly in the outer part of La-MCM-41 and the diameters of the rods were much larger compared to the pore size of La-MCM-41. Moreover, the P/La ratio and the P adsorption capacity are both lower in the case of La_{40}MCM-41 (0.767 $mmol_P\ g^{-1}$) compare to La_{40}SBA-15. This finding showed the significant influence of pore size on the P adsorption behaviours. The results suggested that large pore La-SBA-15 favoured a high accessibility of La components and a fast adsorption of P species within the pore channels. The chemical reaction between P and La lead to a heterogeneous nucleation of $LaPO_4$ species in the confined nanopores. The further growth would result in $LaPO_4$ polycrystalline nanorods (Figure 3.3c), thus achieving high phosphates adsorption capability. As a proof of concept for the practical application, results of blue green algae cultivation showed that La-SBA-15 with high phosphate removal capacity can prevent the excessive plant growth (inset of Figure 3.3b).

In the case of La-SBA-15, it has been demonstrated that the synergistic effect of high La species loading and open porous structure on the phosphate adsorption property. On the basis of this study, a simply hypothesis is that a support material with much larger pore size than of SBA would eliminate the pore blocking issue, thus generating new adsorbents with even higher P adsorption capacity as well as high La usage efficiency. With this consideration, Yang and co-workers further utilized much larger pore support material to confine the phosphate in the pores (Figure 3.4). Macroporous siliceous foam (MOSF) with the pore size of ~100 nm was used as support material (Jie Yang *et al.* 2012). MOSF was synthesized via a supra-assembly approach using cooperatively self-assembled vesicles as building blocks (Wang *et al.* 2006; Yuan *et al.* 2009). Lanthanum oxide as the functional component, derived from $La(NO_3)_3 \cdot 6H_2O$, was incorporated into MOSF using the ethanol evaporation method. MOSF was impregnated with mass ratio of 25 to 200% of $La(NO_3)_3 \cdot 6H_2O$. La_2O_3 component in the functionalized composites was amorphous and homogenously dispersed on the silica surface. TEM observation showed that La_2O_3 layers sandwiched the MOSF walls. No evidence of nanocrystalline La components can be found in the composite material. Selected area electron diffraction (SAED) pattern only indicated that silica walls with La component layers were amorphous. After La impregnation, it was found that the surface area and pore volume of La_xMOSF (x donates the mass ratio between MOSF and La) consistently decreased. This decrease can be attributed to the incorporation of La_2O_3 (with relatively higher density of ~6.5 cm^3 g^{-1}) within the SiO_2 matrix (with a relatively lower density of ~2.2 cm^3 g^{-1}).

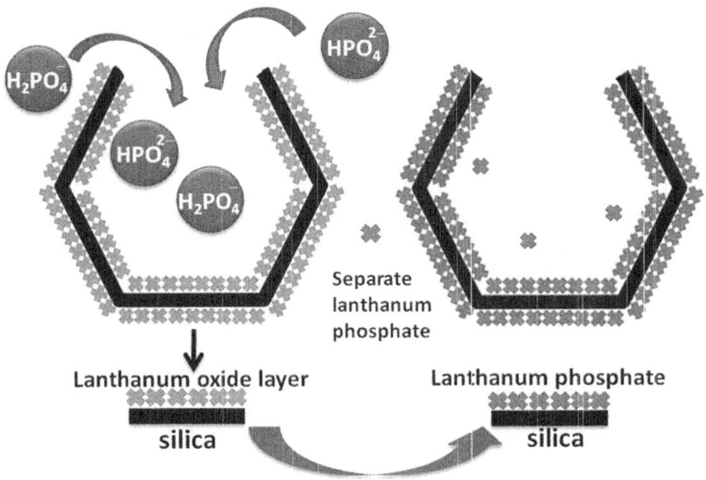

Figure 3.4 Schematic illustration of the P adsorption mechanism. The left shows the typical sandwich-like structure of La–MOSF materials and the right shows the structure of La–MOSF after phosphate adsorption. The pore size of La–MOSF, the diameter of phosphate ions and lanthanum phosphate, the thickness of the lanthanum oxide layer and the lanthanum phosphate layer are not to scale. (Jie Yang *et al.* 2012) Copyright 2012 The Royal Society of Chemistry.

When phosphate adsorption was carried out, MSOF without La functionalization adsorbed a negligible amount of phosphate. In contrast, the La_x MOSF materials showed noticeably higher adsorption capacities which were in line with the amount of impregnated La in the La_x MOSF composites i.e. 0.551, 0.971, 1.874, 2.272 $mmol_P$ g^{-1} ($x = 25$, 50, 100, and 200, respectively). From the kinetic study, $La_{200}MOSF$ achieved equilibrium and phosphorus uptake as high as 1.75 $mmol_P$ g^{-1} within first 200 min, which was faster and higher compared with previous works reported above. Additionally, the rate constant of $La_{200}MOSF$ was calculated to be 0.021 g $mmol_P^{-1}$ min^{-1} which is several times higher than those of other phosphorus adsorbents reported in previous studies (Li *et al.* 2009; Shin *et al.* 2004), indicating that $La_{200}MOSF$ has a ultrafast P removal ability. Meanwhile isotherm study revealed that the adsorption process occurred on the homogeneous surface of La_xMOSF materials through chemical interaction which was in agreement with the observation of La-modified bentonite (Haghseresht *et al.* 2009) and La doped vesuvianite (Li *et al.* 2009). Compared with the Yang's previous work on La doped SBA-15 explained above, at the same La loading amount, the ratio of P to La and Q_{max} of $La_{100}MOSF$ were 1.117 and 1.874 $mmol_P$ g^{-1} , respectively which were higher than those of $La_{100}SBA$-15 (0.812 and 1.362 $mmol_P$ g^{-1}). The higher P/La ratio indicated that the pore blocking problem in the case of La-SBA-15 material at high La loading amount can be tackled by using a large macropore material such as MOSF for the phosphate adsorption application.

To investigate the potential application, P adsorption at low phosphorus concentration range was performed using $La_{200}MOSF$. Referred to maximum discharge limit for phosphorus in wastewater (Ugurlu & Salman, 1998), the concentration of 1 mg L^{-1} is chosen. It showed that after 24 h adsorption, the phosphorus concentration can be decreased from 1 mg L^{-1} to 0.04 mg L^{-1}, indicating that 96% of phosphorus has been removed. Moreover, at low concentration it was found that kinetic study was fitted the pseudo second order kinetic model and the rate constant was 18.67 g $mmol_P^{-1}$ min^{-1}, suggesting that the phosphorus removal rate of $La_{200}MOSF$ was very fast at even a low concentration. The adsorption process can be completed within 30 min. In addition to that, $La_{200}MOSF$ was used to adsorb phosphorus contained in synthetic wastewater after biological treatment. After biological treatment, the concentration of phosphorus was reduced to 5.00 mg L^{-1}, still above the discharge limit of phosphorus. When $La_{200}MOSF$ was applied to that wastewater, the phosphorus concentration was dramatically decreased to 0.025 mg L^{-1} demonstrating the promising potential of $La_{200}MOSF$ to reduce the phosphorus concentration lower than the discharge limit.

To provide direct evidence to support the hypothesis that large pore of MOSF eliminates the pore blocking issue, TEM was employed to study the structure of $La_{200}MOSF$ after adsorption. After 24 hours adsorption, only foam-like macroporous structure can be observed in low-magnification TEM image (Figure 3.5a), indicating that the resultant $LaPO_4$ species were all confined inside the MOSF, thus minimizing the leakage problem of phosphorus. In the dark field-STEM image (Figure 3.5b), nanoparticles with brighter contrast were embedded in the macroporous silica matrix with darker contrast. In the bright-field TEM image (Figure 3.5c) recorded in the same area, some separate dark dots within the pores of MOSF can be observed, further implying the confinement of $LaPO_4$ within MOSF. At the high magnification TEM image (Figure 3.5d), $LaPO_4$ species with two different morphologies can be observed more clearly. On the wall surface, an ordered lattice fringe with a thickness of around 2.5–3.5 nm (indicated by white arrows) can be found. The crystal lattice can be indexed to d_{120} with the spacing of 0.31 nm, suggesting the polycrystalline $LaPO_4$ specie grown on the silica wall (inset of Figure 5d down left). The deep contrast dot-like $LaPO_4$ species indicated by black arrows can also be observed. From the high-resolution TEM image (inset of

Figure 5d upper right), polycrystalline phase with a d_{120} spacing of 0.31 nm were also observed, confirming that such dot-like components are also $LaPO_4$ species.

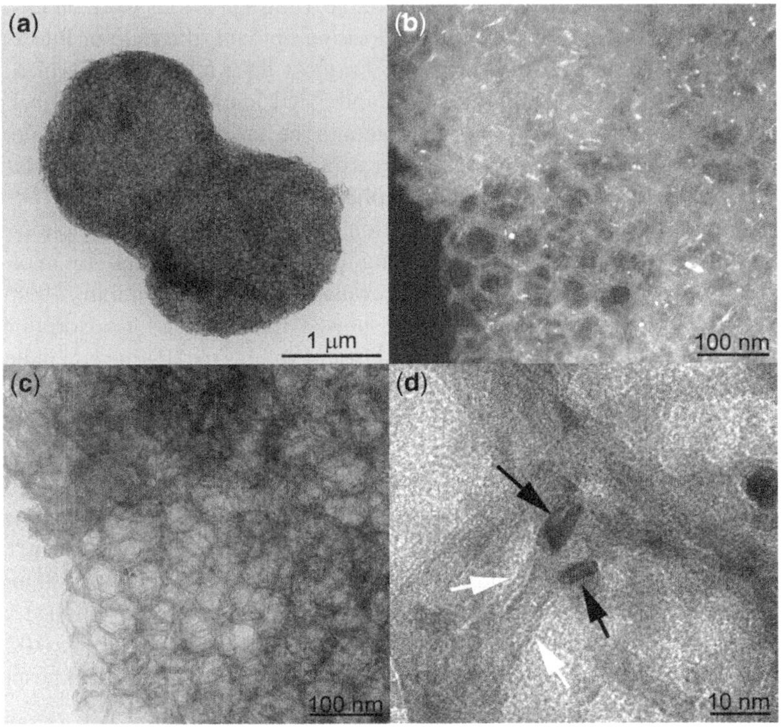

Figure 3.5 (a) TEM image of La_{200}MOSF after adsorption with low magnification; (b) and (c) are DF-STEM image and bright field TEM image of La_{200}MOSF after adsorption; (d) the HRTEM image for La_{200}MOSFafter adsorption. Inset of 9d0 down left and upper right show the high-resolution TEM images of LaPO4 species on the wall surface of MOSF and dot-like LaPO4 species, respectively. (Jie Yang *et al.* 2012) Copyright 2012 The Royal Society of Chemistry.

The superior performance of La_xMOSF for phosphorus removal at both high and low concentration solution was due to the large open pore structure of MSOF which increased both La loading content and La usage efficiency. In the case of La-SBA-15, as explained above, its pore channel could be blocked by the formed $LaPO_4$ nanorods, which caused the reduction of La usage efficiency. The pore block would also made it difficult to further increase the adsorption amount of phosphorus by increasing the impregnation amount of La components. On the contrary, in the La-MOSF case, the large size pores could avoid the blockage occurred during adsorption process and assure the smooth diffusion of phosphate ions into the pores to further react with the La layers. Thus, the loading amount of functional components can be further increased to enhance the adsorption capability without significant reduction in the La usage efficiency. Furthermore, the thin layer of La on the silica walls surface of MOSF resulted in the fast kinetics and excellent performance in phosphorus adsorption at the low concentration range due to the high accessible surface of lanthanum oxide layers to react with phosphate.

From Yang and co-workers' study, it can be concluded that large pore size, open pore network and high pore volume are essential structural features for adsorbents with outstanding adsorption performance. However, there are at least two issues that remained unclear i.e. the influences of mesostructured on the growth behaviours of confined $LaPO_4$, and the consequent impact on phosphate adsorption properties. To answer these questions, they further chose an ordered three-dimensional (3D) cage-type mesoporous silica material, FDU-12, as the porous support matrix (Yang *et al.* 2013). In the mesostructured of FDU-12, the cage-like pores were interconnected by entrances and the entrance size could be fine adjusted by controlling synthesis conditions. The general concept of this study was investigating the influences of different entrance size on the growth behaviour of $LaPO_4$ and evaluating the overall impact on phosphate adsorption performance. FDU-12 materials were synthesized according to the method reported by Fan et.al with the minor modification (pH adjustment before hydrothermal treatment) (Fan *et al.* 2005). Different hydrothermal treatment temperatures i.e. 100°C, 120°C, and 180°C were employed in order to adjust the entrance size of FDU-12. The fine control of entrance sizes was indicated by nitrogen-sorption analysis. The sizes can be adjusted from 4.6 to 5.8 nm. FDU-12 materials were further functionalized with different amounts of lanthanum oxide (La_x-FDU-T, x was the mass ratio as a percentage of La $(NO_3)_3 \cdot 6H_2O$ versus FDU-12, ($x = 25$, 50, 75, 100, 150 and 200); T denotes the hydrothermal treatment temperature).

Testing on phosphate adsorption performance was carried out in all La_xFDU-12-T ($x = 25$, 50, 75, 100, 150 and 200 and T = 100, 160, 180°C) samples. For La_xFDU-12–100, the Q_{max} that can be reached is 0.632 $mmol_P\ g^{-1}$ at La loading ratio is 50%. However with the increase of impregnated La concentration, the molar ratio of P adsorbed versus La (defined as P/La) decreased to 0.650 indicating that the majority of La_2O_3 in La_xFDU-12–100 cannot be fully utilized. Meanwhile for other samples, with a relatively low La loading amount ($25 \leq x \leq 100$), almost all La species in La_xFDU-12–160 and La_xFDU-12–180 can be utilized for phosphate adsorption, as indicated by the P/La value of approximately 1. When higher La loading amount ($150 \leq x \leq 200$) was used, the P/La value decreased, which indicated that there was a slight reduction in the La usage efficiency. The highest Q_{max} that can be achieved among all samples was 2.224 $mmol_P\ g^{-1}$. In addition to that, from kinetic study it revealed that the adsorption process of phosphate in La_xFDU-12-T was close to chemisorption and the kinetic rate constants demonstrated that La_xFDU-12-T with larger entrance can adsorb phosphate faster. As an example, La_{25}FDU-12–180 shows a high rate constant (k) of 0.173 $g mmol^{-1} min^{-1}$, whereas

La_{25}FDU-12–100 shows a much smaller k of only 0.036 $g mmol^{-1} min^{-1}$, clearly demonstrating that La_x FDU-12-T with larger entrance size can adsorb phosphate faster.

To understand the influence of entrance size and La loading content on the phosphate adsorption performances, further characterisations on P adsorbed La_xFDU-12-T were carried out. The result from 3D reconstructed Electron Tomography (ET) image showed that depending on the entrance size, chemisorbed phosphate can be confined either in isolated cages or in both cages and entrances (Figure 3.6a–e). In the case of La_{100}FDU-12–100, the $LaPO_4$ nano-rod (represented by red in Figure 3.6a) was strictly confined in an individual cage (shown by the green in Figure 3.6a) and the length of the nano-rod was restricted by the size of the cage. This observation was further confirmed by reconstructed 3D structure (Figure 3.6b), implying that crystalline $LaPO_4$ can only grow inside the cages instead of inside the entrances. Therefore it was difficult to fully utilise the lanthanum oxide in the entrances for phosphate adsorption, and chemisorbed phosphates were only confined in separated cages, which would result in the poor La usage efficiency and low adsorption performance. For comparison, La_{100}FDU-12–180 with larger entrance size was characterised. After phosphate adsorption, a rodlike $LaPO_4$ crystal with a length of approximately 75 nm (indicated in red) in the connected cages can be clearly observed in ET slice image (Figure 3.6c). The reconstructed 3D structures of long $LaPO_4$ nano-rod shown in Figure 3.6d and e demonstrated the branchlike structural detail, suggesting their growth direction followed the direction of the cages connections. Such crystalline growth behaviour would enable the occupation of spaces in both the entrances and cages. For FDU-12-T materials with the same La loading content, La_xFDU-12–180 with largest entrance size showed a faster adsorption rate because the blocking of the entrances by $LaPO_4$ was not as severe as the other samples with small entrance sizes. Phosphate ions can diffuse into the porous networks more easily through the entrances, thus leading to a fast adsorption rate. For La_xFDU-12–180 with different La loading amounts, the extent of the entrance blocking was dependent on the La loading content. Materials with higher x can adsorb more phosphates and generate more resultant $LaPO_4$, thereby causing the entrance block and the low adsorption rate.

In Yang and co-workers' series reports, it has demonstrated that the phosphate adsorption performance can be governed by controlling the nucleation and growth of $LaPO_4$ inside the porous entrances/cages using adsorbents with sufficiently large pore diameters. Owning to the growth of $LaPO_4$ nanocrystals inside the pores, the relatively small mesopores of La-SBA-15 and La-FDU-12 restrained the high loading content of active La components, thus the phosphate adsorption capacity and the La usage efficiency. La-MOSF materials with pore size larger than 50 nm can effectively eliminate the pore blocking problem and realize superior phosphate adsorption properties. However, all the aforementioned nanoporous silica materials (SBA-15, FDU-12, and MOSF) are all prepared by surfactant template synthesis method, followed by a wet impregnation process. The complicated, high cost and time consuming procedure of this protocol is a major drawback for its extensive industrial scale application. For this consideration, an efficient, facile and cost efficient method for the synthesis of alternative high performance adsorbents is highly desirable. From this point of view, Emmanuelawati and co-workers reported a facile spray drying method for the large-scale fabrication of La doped highly porous silica microspheres (Emmanuelawati *et al.* 2013). Using this method, La can be introduced in situ during spray drying process and converted into lanthanum oxide by post-calcination process, avoiding the impregnation process conventionally used to incorporate La. The resultant product from this method has appealing properties such as large size open pores and high pore volumes, which were favourable for excellent phosphate adsorption performance and fast kinetic at high or even low phosphate concentrations. Aerosil 200, known as commercialized hydrophilic fumed silica, was used as the sole precursor of porous silica microspheres. Further, $La(NO_3)\cdot 6H_2O$ as La precursor, was introduced in situ with the mass ratio percentage of 100, 150 and 200%. After spray drying and calcination, Aerosil 200 was assembled into microspherical granules size of 1–5 μm. From SEM observations, large inter-particle voids can be directly found. N_2 adsorption-desorption result gave a broad pore size distribution centred at 65 nm. Large pore volume could be generated due to the formation of large voids during spray drying process. The La functionalized aerosil 200 microspheres retained similar morphology after functionalization. In addition to that, the N_2 adsorption-desorption of La functionalized aerosol 200 confirmed the existence of macropore or large mesopore. According to previous studies, large mesopore and macropore were beneficial over small mesopore with pore size less than 10 nm in phosphate adsorption (Jie Yang *et al.* 2012; Yang *et al.* 2013; Jie Yang *et al.* 2011). From adsorption isotherm study, there was an indication that the adsorption of phosphate ions was taking place in a multilayer adsorption fashion as all adsorption isotherm data followed the Freundlich model. Moreover, n value of The Freundlich isotherm was between 2 and 10, suggesting that the phosphate anions can be easily adsorbed. The maximum phosphate adsorption capacities for La functionalized aerosol were 1.946, 2.317, 2.242 mmol g^{-1} for 100, 150 and 200% La-loading materials, respectively. At low concentration of phosphate of 1 mg L^{-1}, 150 % La-loading material can remove 94% of phosphate ion within 24 hours. Kinetic study showed the rate constant value was higher than that of La_{200}MOSF reported previously (Jie Yang *et al.* 2012), suggesting the fast adsorption rate of La functionalized at a low P concentration. Based on elemental mapping, XRD and TEM results of the products after adsorption, it was concluded that homogenously distributed La_2O_3 inside the aerosol microspheres together with the large pores and high pore volumes lead to the fast adsorption of phosphate ions bound to the active sites and the eventual growth of $LaPO_4$ nanocrystals confined within the La functionalized aerosol microspheres.

3.8 CRITICAL RESEARCH NEEDS

The coming years will see the further development of adsorption towards a mature technology in the water treatment application. Nanoporous materials have shown their great potential in laboratory study. Their future development and commercialization still face a variety of challenges.

On the materials side, further research need to identify the most appropriate pore size distribution. Large pore size nanoporous materials evidenced with remarkable performance usually have relatively low surface area and excess intraparticle voids, which is against the common notion that high surface area is essential for adsorption. The excess voids within the adsorbents would could a low volume adsorption performance. Flow transport limitation is another problem commonly encountered in nanoporous materials. Thus, we need to consider the porosity and pore architecture of entire adsorption column rather than only the pore characteristics of individual nanoporous particles to arrive at a adsorption technology that combines a high adsorption capacity with high removal efficiency.

The choice of nanoporous silica support materials for adsorbents, as we have discussed, is well substantiated by the high P and As removal performances. However, it worth noting that in principle other materials, such as carbons, polymers, and ceramics, can also be used as support subtracts as long as they form a desired nanoporous network. Advanced engineering of such materials would potentially enable the fabrication of multi-functional adsorbents with high overall water purification properties.

The long-term performance and the re-cycle of adsorbent materials should be addressed to meet the realistic application requirements. So far, most laboratory studies are only carried out for short period of time. The recycle studies of nanoporous adsorbents are very rare. Furthermore, considering the complicity of natural water and practical wastewater, assessment on the applicability and efficiency of nanoporous material based adsorbents should be carried out under more realistic conditions. Accordingly, the stability of both nanoporous adsorbent itself and the spent materials should be evaluated under similar condition to prevent the possible risk involved.

Economic consideration is a key importance issue to make nanoporous material adsorption an affordable mass-market technology. The current fabrication cost of nanoporous adsorbents remains high. Hence, the investigation on how to reduce cost issue and finding alternative trade-off is necessary.

3.9 CONCLUSION

Water is an important part of human life, affecting every aspect such as human health, food production, energy, industrial activity as well as the quality of environment. Pollutants, such as arsenic and phosphate, are the major pollutants that can found in water. The presence of these pollutants is mainly due to the natural process and the increasing activities in industry and agriculture. Adsorption has been considered as an effective and efficient route to remove arsenic and phosphate in water. Adsorption offers flexibility in design and operation, ability to produce the high quality effluent and also relatively low cost-technology compared to other technologies for arsenic and phosphate removal. Successful application of adsorption technology in removal pollutant such as arsenic and phosphate requires the development of high performance adsorbents which can achieve high removal capacity, high adsorption rate and efficient cost. As explained above, the rational design of nanoporous adsorbent to confine arsenic and phosphate includes the control of pores structure of the adsorbent and the proper incorporation of active materials (iron oxide for arsenic and lanthanum oxide for phosphate adsorption) into the supporting matrix such as carbon and silica to provide active sites for adsorption to take place.

The control of pore structures can be achieved by fine controlling of the synthesis conditions. In the cases of arsenic and phosphate adsorption, there are fundamental correlations between pore structure and adsorption performance. Pore sizes of the adsorbents after functionalization with iron oxide and lanthanum oxide directly determine the confining behaviours of both arsenic and phosphate species. The large open porous network can promote the fast diffusion/ transportation of solutions in the adsorption process. For phosphate adsorption, the heterogeneous nucleation of $LaPO_4$ generated from chemical reaction between P and La would be confined in the porous network. Large pores would also provide advantages in tackling the issue related to the pore blocking problem caused by the nucleation and growth of $LaPO_4$ crystals. In addition to that, the right choice of incorporated active materials and their distribution in the adsorbent matrix would also influence the performance of the adsorbents. Aggregation prevented active materials such as iron oxide and lanthanum oxide along with rationally designed pore structures, will provide numerous active sites for adsorption, achieving high adsorption capacities and fast adsorption rates.

Despite the significant progresses achieved in this field, some issues and challenges remain in the material basis, performance evaluation and manufacturing costs. The research for improved adsorbent materials and overall technical solutions to enhance adsorption performance continues.

3.10 REFERENCES

Abernathy C. O., Liu Y. P., Longfellow D., Aposhian H. V., Beck B., Fowler B. and Waalkes M. (1999). Arsenic: health effects, mechanisms of actions, and research issues. *Environmental Health Perspectives*, **107**(7), 593–597.

Amin M. T., Alazba A. A. and Manzoor U. (2014). A review of removal of pollutants from water/wastewater using different types of nanomaterials. *Advances in Materials Science and Engineering*, **2014**, 24. doi: 10.1155/2014/825910

Andjelkovic I., Tran D. N. H., Kabiri S., Azari S., Markovic M. and Losic D. (2015). Graphene aerogels decorated with α-FeOOH nanoparticles for efficient adsorption of arsenic from contaminated waters. *Acs Applied Materials and Interfaces*, **7**(18), 9758–9766. doi: 10.1021/acsami.5b01624

Argos M., Kalra T., Rathouz P. J., Chen Y., Pierce B., Parvez F. and Ahsan H. (2010). Arsenic exposure from drinking water, and all-cause and chronic-disease mortalities in Bangladesh (HEALS): a prospective cohort study. *The Lancet*, **376**(9737), 252–258. doi: http://dx.doi.org/10.1016/S0140-6736(10)60481-3

Baikousi M., Bourlinos A. B., Douvalis A., Bakas T., Anagnostopoulos D. F., Tuček J. and Karakassides M. A. (2012). Synthesis and characterization of γ-Fe2O3/Carbon hybrids and their application in removal of hexavalent chromium ions from aqueous solutions. *Langmuir*, **28**(8), 3918–3930. doi: 10.1021/la204006d

Barrett J. R. (2014). Chemical Contaminants in drinking water: where do we go from here? *Environmental Health Perspectives*, **122**(3), A80–A80. doi: 10.1289/ehp.122-A80

Bissen M. and Frimmel F. H. (2003). Arsenic—a review. Part II: Oxidation of arsenic and its removal in water treatment. *Acta hydrochimica et hydrobiologica*, **31**(2), 97–107. doi: 10.1002/aheh.200300485.

Biswas B. K., Inoue K., Ghimire K. N., Ohta S., Harada H., Ohto K. and Kawakita H. (2007). The adsorption of phosphate from an aquatic environment using metal-loaded orange waste. *Journal of Colloid and Interface Science*, **312**(2), 214–223. doi: http://dx.doi.org/10.1016/j.jcis.2007.03.072

Borho M. and Wilderer, P. (1996). Optimized removal of arsenate(III) by adaptation of oxidation and precipitation processes to the filtration step. *Water Science and Technology*, **34**(9), 25–31. doi: 10.1016/S0273-1223(96)00783-4

Burke A. M., Hanrahan J. P., Healy D. A., Sodeau J. R., Holmes J. D. and Morris M. A. (2009). Large pore bi-functionalised mesoporous silica for metal ion pollution treatment. *Journal of Hazardous Materials*, **164**(1), 229–234. doi: http://dx.doi.org/10.1016/j.jhazmat.2008.07.146

Calderon R. L. (2000). The epidemiology of chemical contaminants of drinking water. *Food and Chemical Toxicology*, **38**, S13–S20. doi: 10.1016/s0278-6915(99)00133-7

Cao C.-Y., Li P., Qu J., Dou Z.-F., Yan W.-S., Zhu J.-F. and Song W.-G. (2012). High adsorption capacity and the key role of carbonate groups for heavy metal ion removal by basic aluminum carbonate porous nanospheres. *Journal of Materials Chemistry*, **22**(37), 19898–19903. doi: 10.1039/c2jm34138g

Chandra V., Park J., Chun Y., Lee J. W., Hwang I.-C. and Kim K. S. (2010). Water-dispersible magnetite-reduced graphene oxide composites for arsenic removal. *ACS Nano*, **4**(7), 3979–3986. doi: 10.1021/nn1008897

Chen W., Parette R., Zou J., Cannon F. S. and Dempsey B. A. (2007). Arsenic removal by iron-modified activated carbon. *Water Research*, **41**(9), 1851–1858. doi: http://dx.doi.org/10.1016/j.watres.2007.01.052

Davis M. E. (2002). Ordered porous materials for emerging applications. *Nature*, **417**(6891), 813–821.

Delaney P., McManamon C., Hanrahan J. P., Copley M. P., Holmes J. D. and Morris M. A. (2011). Development of chemically engineered porous metal oxides for phosphate removal. *Journal of Hazardous Materials*, **185**(1), 382–391.

Elcik H., Cakmakci M., Sahinkaya E. and Ozkaya B. (2013). Arsenic removal from drinking water using low pressure membranes. *Industrial and Engineering Chemistry Research*, **52**(29), 9958–9964. doi: 10.1021/ie401393p

Emmanuelawati I., Yang J., Zhang J., Zhang H., Zhou L. and Yu, C. (2013). Low-cost and large-scale synthesis of functional porous materials for phosphate removal with high performance. *Nanoscale*, **5**(13), 6173–6180. doi: 10.1039/C3NR01574B

Falconer I. R. (2005). Health risk assessment of cyanobacterial (Blue-green Algal) toxins in drinking water. *International Journal of Environmental Research and Public Health*, **2**(1), 43–50. doi: 10.3390/ijerph2005010043

Fan J., Yu C., Lei J., Zhang Q., Li T., Tu B. and Zhao D. (2005). Low-temperature strategy to synthesize highly ordered mesoporous silicas with very large pores. *Journal of the American Chemical Society*, **127**(31), 10794–10795. doi: 10.1021/ja052619c

Fawell J. and Nieuwenhuijsen M. J. (2003). Contaminants in drinking water: environmental pollution and health. *British Medical Bulletin*, **68**(1), 199–208. doi: 10.1093/bmb/ldg027

Fontcuberta M., Calderon J., Villalbí J. R., Centrich F., Portaña S., Espelt A. and Nebot, M. (2011). Total and inorganic arsenic in marketed food and associated health risks for the catalan (Spain) population. *Journal of Agricultural and Food Chemistry*, **59**(18), 10013–10022. doi: 10.1021/jf2013502

Frazer L. (2005). Metal attraction: an ironclad solution to arsenic contamination? *Environmental Health Perspectives*, **113**(6), A398–A401.

Genç Fuhrman H. (2004). Adsorption of arsenic from water using activated neutralized red mud. *Environmental Science and Technology*, **38**(8), 2428–2434. doi: 10.1021/es035207h

Georgakilas V., Otyepka M., Bourlinos A. B., Chandra V., Kim N., Kemp K. C. and Kim K. S. (2012). Functionalization of graphene: covalent and non-covalent approaches, derivatives and applications. *Chemical Reviews*, **112**(11), 6156–6214. doi: 10.1021/cr3000412

Georgantas D. A. and Grigoropoulou H. P. (2007). Orthophosphate and metaphosphate ion removal from aqueous solution using alum and aluminum hydroxide. *Journal of Colloid and Interface Science*, **315**(1), 70–79. doi: http://dx.doi.org/10.1016/j.jcis.2007.06.058

Gracia-Espino E., López-Urías F., Terrones H. and Terrones M. (2012). Chapter 1 - Novel nanocarbons for adsorption. In: Novel Carbon Adsorbents, J. M. D. Tascón (ed.), 2nd edn, Elsevier, Oxford, pp. 3–34.

Gu Z., Fang J. and Deng B. (2005). Preparation and evaluation of GAC-based iron-containing adsorbents for arsenic removal. *Environmental Science and Technology*, **39**(10), 3833–3843. doi: 10.1021/es048179r

Gu Z., Deng B. and Yang J. (2007). Synthesis and evaluation of iron-containing ordered mesoporous carbon (FeOMC) for arsenic adsorption. *Microporous and Mesoporous Materials*, **102**(1–3), 265–273. doi: http://dx.doi.org/10.1016/j.micromeso.2007.01.011

Haghseresht F., Wang S. and Do D. D. (2009). A novel lanthanum-modified bentonite, Phoslock, for phosphate removal from wastewaters. *Applied Clay Science*, **46**(4), 369–375. doi: http://dx.doi.org/10.1016/j.clay.2009.09.009

Ho Y. S. and McKay G. (2000). The kinetics of sorption of divalent metal ions onto sphagnum moss peat. *Water Research*, **34**(3), 735–742. doi: http://dx.doi.org/10.1016/S0043-1354(99)00232-8

Hosni K., Ben Moussa S. and Ben Amor M. (2007). Conditions influencing the removal of phosphate from synthetic wastewater: influence of the ionic composition. *Desalination*, **206**(1–3), 279–285.

Huang W. Y., Zhu Y., Tang J. P., Yu X., Wang X. L., Li D. and Zhang Y. M. (2014). Lanthanum-doped ordered mesoporous hollow silica spheres as novel adsorbents for efficient phosphate removal. *Journal of Materials Chemistry A*, **2**(23), 8839–8848.

Jang M., Chen W. and Cannon F. S. (2008). Preloading hydrous ferric oxide into granular activated carbon for arsenic removal. *Environmental Science and Technology*, **42**(9), 3369–3374. doi: 10.1021/es7025399

Jekel M., Genz A. and Stindt U. (2002). Studies on Phosphorus Removal from Fresh and Sea Water by Commercial Sorbents. TU Berlin, Berlin.

Jiang F., Beck M. B., Cummings R. G., Rowles K. and Russell D. (2004). Estimation of Costs of Phosphorus Removal in Wastewater Treatment Facilities: Construction De Novo. (#2004–010). Georgia Water Planning and Policy Center, Andrew Young School of Policy Studies, Georgia State University, Atlanta.

Jiang J. Q. (2001). Removing arsenic from groundwater for the developing world: a review. *Water Science and Technology*, **44**(6), 89–98.

Kanel S. R. and Nepal D. (2009). 11. Nanoscale porous materials for water treatment: advances and challenges. In: Nanotechnologies for Water Environment Applications: American Society of Civil Engineers (ASCE), T. C. Zhang, R. Y. Surampalli, K. C. K. Lai, Z. Hu, R. D. Tyagi and I. M. C. Lo (eds.). American Society of Civil Engineers, Virginia, USA, doi: 10.1061/9780784410301.ch11

Katsoyiannis I. A. and Zouboulis A. I. (2002). Removal of arsenic from contaminated water sources by sorption onto iron-oxide-coated polymeric materials. *Water Research*, **36**(20), 5141–5155. doi: http://dx.doi.org/10.1016/S0043-1354(02)00236-1

Kim K. S., Zhao Y., Jang H., Lee S. Y., Kim J. M., Kim K. S. and Hong B. H. (2009). Large-scale pattern growth of graphene films for stretchable transparent electrodes. *Nature*, **457**(7230), 706–710.

Len Ritter K. S., Sibley P., Hall K., Patricia Keen G. M. and Beth L. (2002). Sources, pathways, and relative risks of contaminants in surface water and groundwater: a perspective prepared for the walkerton inquiry. *Journal of Toxicology and Environmental Health, Part A*, **65**(1), 1–142. doi: 10.1080/152873902753338572

Li H., Ru J., Yin W., Liu X., Wang J. and Zhang W. (2009). Removal of phosphate from polluted water by lanthanum doped vesuvianite. *Journal of Hazardous Materials*, **168**(1), 326–330. doi: http://dx.doi.org/10.1016/j.jhazmat.2009.02.025

Lima A. A. M., Moore S. R., Barboza M. S., Soares A. M., Schleupner M. A., Newman R. D. and Guerrant R. L. (2000). Persistent diarrhea signals a critical period of increased diarrhea burdens and nutritional shortfalls: a prospective cohort study among children in Northeastern Brazil. *Journal of Infectious Diseases*, **181**(5), 1643–1651. doi: 10.1086/315423

Liu R., Shi Y., Wan Y., Meng Y., Zhang F., Gu D. and Zhao D. (2006). Triconstituent co-assembly to ordered mesostructured polymer – silica and carbon – silica nanocomposites and large-pore mesoporous carbons with high surface areas. *Journal of the American Chemical Society*, **128**(35), 11652–11662. doi: 10.1021/ja0633518

Lu A.-H., Nitz J.-J., Comotti M., Weidenthaler C., Schlichte K., Lehmann C. W. and Schüth F. (2010). Spatially and size selective synthesis of fe-based nanoparticles on ordered mesoporous supports as highly active and stable catalysts for ammonia decomposition. *Journal of the American Chemical Society*, **132**(40), 14152–14162. doi: 10.1021/ja105308e

Ma J., Zhu Z., Chen B., Yang M., Zhou H., Li C. and Chen J. (2013). One-pot, large-scale synthesis of magnetic activated carbon nanotubes and their applications for arsenic removal. *Journal of Materials Chemistry A*, **1**(15), 4662–4666. doi: 10.1039/C3TA10329C

Meiser F., Cortez C. and Caruso F. (2004). Biofunctionalization of fluorescent rare-earth-doped lanthanum phosphate colloidal nanoparticles. *Angewandte Chemie International Edition*, **43**(44), 5954–5957. doi: 10.1002/anie.200460856

Mohan D. and Pittman Jr, C. U. (2007). Arsenic removal from water/wastewater using adsorbents—A critical review. *Journal of Hazardous Materials*, **142**(1–2), 1–53. doi: http://dx.doi.org/10.1016/j.jhazmat.2007.01.006

Mondal P., Majumder C. B. and Mohanty B. (2006). Laboratory based approaches for arsenic remediation from contaminated water: Recent developments. *Journal of Hazardous Materials*, **137**(1), 464–479. doi: http://dx.doi.org/10.1016/j.jhazmat.2006.02.023

Morse G. K., Brett S. W., Guy J. A. and Lester J. N. (1998). Review: Phosphorus removal and recovery technologies. *Science of the Total Environment*, **212**(1), 69–81. doi: http://dx.doi.org/10.1016/S0048-9697(97)00332-X

Mou F., Guan J., Xiao Z., Sun Z., Shi W. and Fan X.-A. (2011). Solvent-mediated synthesis of magnetic Fe2O3 chestnut-like amorphous-core/[gamma]-phase-shell hierarchical nanostructures with strong As(v) removal capability. *Journal of Materials Chemistry*, **21**(14), 5414–5421. doi: 10.1039/c0jm03726e

Ning R. Y. (2002). Arsenic removal by reverse osmosis. *Desalination*, **143**(3), 237–241. doi: http://dx.doi.org/10.1016/S0011-9164(02)00262-X

O'Connor A. J., Hokura A., Kisler J. M., Shimazu S., Stevens G. W. and Komatsu, Y. (2006). Amino acid adsorption onto mesoporous silica molecular sieves. *Separation and Purification Technology*, **48**(2), 197–201. doi: http://dx.doi.org/10.1016/j.seppur.2005.07.007

Ou E., Junjie Z. J., Shaochun M. C., Jiaqiang W. Q., Fei X. and Liang, M. (2007). Highly efficient removal of phosphate by lanthanum-doped mesoporous SiO2. *Colloids and Surfaces a-Physicochemical and Engineering Aspects*, **308**(1–3), 47–53.

Qu X., Alvarez P. J. J. and Li Q. (2013). Applications of nanotechnology in water and wastewater treatment. *Water Research*, **47**(12), 3931–3946. doi: http://dx.doi.org/10.1016/j.watres.2012.09.058

Ravenscroft P., Brammer H. and Richards K. (2009). Arsenic Pollution: A Global Synthesis (Vol. 28). John Wiley and Sons, West Sussex, United Kingdom.

Reardon R. (2006). Technical introduction of membrane separation processes for low TP limits. Paper presented at the WERF, ALexandria.

Rodriguez-Mozaz S., López de Alda M. J. and Barceló D. (2004). Monitoring of estrogens, pesticides and bisphenol A in natural waters and drinking water treatment plants by solid-phase extraction–liquid chromatography–mass spectrometry. *Journal of Chromatography A*, **1045**(1–2), 85–92. doi: http://dx.doi.org/10.1016/j.chroma.2004.06.040

Saad R., Hamoudi S. and Belkacemi K. (2008). Adsorption of phosphate and nitrate anions on ammonium-functionnalized mesoporous silicas. *Journal of Porous Materials*, **15**(3), 315–323. doi: 10.1007/s10934-006-9095-x

Sawicki R. and Mercier L. (2006). Evaluation of mesoporous cyclodextrin-silica nanocomposites for the removal of pesticides from aqueous media. *Environmental Science and Technology*, **40**(6), 1978–1983. doi: 10.1021/es051441r

Shannon M. A., Bohn P. W., Elimelech M., Georgiadis J. G., Marinas B. J. and Mayes A. M. (2008). Science and technology for water purification in the coming decades. *Nature*, **452**(7185), 301–310.

Shin E. W., Han J. S., Jang M., Min S.-H., Park J. K. and Rowell R. M. (2004). Phosphate adsorption on aluminum-impregnated mesoporous silicates: surface structure and behavior of adsorbents. *Environmental Science and Technology*, **38**(3), 912–917. doi: 10.1021/es030488e

Singh R., Singh S., Parihar P., Singh V. P. and Prasad S. M. (2015). Arsenic contamination, consequences and remediation techniques: a review. *Ecotoxicology and Environmental Safety*, **112**, 247–270. doi: 10.1016/j.ecoenv.2014.10.009

Sridhar V., Jung J. H. and Oh I. K. (2011). Microwave syntheses of graphene and graphene decorated with metal nanoparticles. *Carbon*, **49**(13), 4449–4457.

Theron J., Walker J. A. and Cloete T. E. (2008). Nanotechnology and water treatment: applications and emerging opportunities. *Critical Reviews in Microbiology*, **34**(1), 43–69. doi: doi:10.1080/10408410701710442

Tian N., Tian X., Ma L., Yang C., Wang Y., Wang Z. and Zhang L. (2015). Well-dispersed magnetic iron oxide nanocrystals on sepiolite nanofibers for arsenic removal. *Rsc Advances*, **5**(32), 25236–25243. doi: 10.1039/C5RA01592H

Tuček J., Kemp K. C., Kim K. S. and Zbořil R. (2014). Iron-oxide-supported nanocarbon in lithium-ion batteries, medical, catalytic, and environmental applications. *ACS Nano*, **8**(8), 7571–7612. doi: 10.1021/nn501836x

Tuutijärvi T., Lu J., Sillanpää M. and Chen G. (2009). As(V) adsorption on maghemite nanoparticles. *Journal of Hazardous Materials*, **166**(2-3), 1415–1420. doi: http://dx.doi.org/10.1016/j.jhazmat.2008.12.069

Ugurlu A. and Salman B. (1998). Phosphorus removal by fly ash. *Environment International*, **24**(8), 911–918. doi: http://dx.doi.org/10.1016/S0160-4120(98)00079-8

Vadahanambi S., Lee S.-H., Kim W.-J. and Oh I.-K. (2013). Arsenic removal from contaminated water using three-dimensional graphene-carbon nanotube-iron oxide nanostructures. *Environmental Science and Technology*, **47**(18), 10510–10517. doi: 10.1021/es401389g

Vaudreuil S., Bousmina M., Kaliaguine S. and Bonneviot L. (2001). Synthesis of macrostructured silica by sedimentation–aggregation. *Advanced Materials*, **13**(17), 1310–1312. doi: 10.1002/1521-4095(200109)13:17 < 1310::AID-ADMA1310 > 3.0.CO;2-A.

Wang H., Zhou X., Yu M., Wang Y., Han L., Zhang J. and Yu C. (2006). Supra-assembly of siliceous vesicles. *Journal of the American Chemical Society*, **128**(50), 15992–15993. doi: 10.1021/ja066707o

WHO (2011). Guidelines for Drinking-Water Quality. World Health Organization, Geneva, Switzerland.

Wu Z. and Zhao D. (2011). Ordered mesoporous materials as adsorbents. *Chemical Communications*, **47**(12), 3332–3338. doi: 10.1039/C0CC04909C

Wu Z., Li W., Webley P. A. and Zhao D. (2012). General and controllable synthesis of novel mesoporous magnetic iron Oxide@Carbon encapsulates for efficient arsenic eemoval. *Advanced Materials*, **24**(4), 485–491. doi: 10.1002/adma.201103789

Xu Z., Jing C., Li F. and Meng X. (2008). Mechanisms of photocatalytical degradation of monomethylarsonic and dimethylarsinic acids using nanocrystalline titanium dioxide. *Environmental Science and Technology*, **42**(7), 2349–2354. doi: 10.1021/es0719677

Yang J., Zhou L., Zhao L., Zhang H., Yin J., Wei G. and Yu C. (2011). A designed nanoporous material for phosphate removal with high efficiency. *Journal of Materials Chemistry*, **21**(8), 2489. doi: 10.1039/c0jm02718a

Yang J., Yuan P., Chen H.-Y., Zou J., Yuan Z. and Yu C. (2012). Rationally designed functional macroporous materials as new adsorbents for efficient phosphorus removal. *Journal of Materials Chemistry*, **22**(19), 9983–9990. doi: 10.1039/C2JM16681J

Yang J., Zhou L., Zhang J., Zou J., Yuan Z. G. and Yu C. Z. (2013). Confinement of chemisorbed phosphates in a controlled nanospace with three-dimensional mesostructures. *Chemistry-a European Journal*, **19**(18), 5578–5585. doi: 10.1002/chem.201300273

Yang J., Zhang H., Yu M., Emmanuelawati I., Zou J., Yuan Z. and Yu C. (2014). High-content, well-dispersed γ-Fe2O3 nanoparticles encapsulated in macroporous silica with superior arsenic removal performance. *Advanced Functional Materials*, **24**(10), 1354–1363. doi: 10.1002/adfm.201302561

Yavuz C. T., Mayo J. T., Yu W. W., Prakash A., Falkner J. C., Yean S. and Colvin V. L. (2006a). Low-field magnetic separation of monodisperse Fe3O4 nanocrystals. *Science*, **314**(5801), 964–967 %R 910.1126/science.1131475

Yavuz C. T., Mayo J. T., Yu W. W., Prakash A., Falkner J. C., Yean S. and Colvin V. L. (2006b). Low-field magnetic separation of monodisperse Fe3O4 nanocrystals. *Science*, **314**(5801), 964–967. doi: 10.1126/science.1131475

Yean S., Cong L., Yavuz C. T., Mayo J. T., Yu W. W., Kan A. T. and Tomson M. B. (2005). Effect of magnetite particle size on adsorption and desorption of arsenite and arsenate. *Journal of Materials Research*, **20**(12), 3255–3264. doi: doi:10.1557/jmr.2005.0403

Yeoman S., Stephenson T., Lester J. N. and Perry R. (1988). The removal of phosphorus during wastewater treatment: A review. *Environmental Pollution*, **49**(3), 183–233. doi: http://dx.doi.org/10.1016/0269-7491(88)90209-6

Yu X., Tong S., Ge M., Zuo J., Cao C. and Song W. (2013). One-step synthesis of magnetic composites of cellulose@iron oxide nanoparticles for arsenic removal. *Journal of Materials Chemistry A*, **1**(3), 959–965. doi: 10.1039/C2TA00315E

Yu Y. and Paul Chen J. (2015). Key factors for optimum performance in phosphate removal from contaminated water by a Fe–Mg–La tri-metal composite sorbent. *Journal of Colloid and Interface Science*, **445**, 303–311. doi: http://dx.doi.org/10.1016/j.jcis.2014.12.056

Yuan P., Zhou X., Wang H., Liu N., Hu Y., Auchterlonie G. J. and Yu C. (2009). Electron-tomography determination of the packing structure of macroporous ordered siliceous foams assembled from vesicles. *Small*, **5**(3), 377–382. doi: 10.1002/smll.200801020

Zhang G., Khorshed A. and Paul Chen J. (2013). Simultaneous removal of arsenate and arsenite by a nanostructured zirconium–manganese binary hydrous oxide: behavior and mechanism. *Journal of Colloid and Interface Science*, **397**(0), 137–143. doi: http://dx.doi.org/10.1016/j.jcis.2012.11.056

Zhang J., Shen Z., Shan W., Chen Z., Mei Z., Lei Y. and Wang W. (2010). Adsorption behavior of phosphate on Lanthanum(III) doped mesoporous silicates material. *Journal of Environmental Sciences*, **22**(4), 507–511. doi: http://dx.doi.org/10.1016/S1001-0742(09)60141-8

Zhang Q., Pan B., Zhang W., Pan B., Zhang Q. and Ren H. (2008). Arsenate removal from aqueous media by nanosized hydrated ferric oxide (HFO)-Loaded polymeric sorbents: effect of HFO loadings. *Industrial and Engineering Chemistry Research*, **47**(11), 3957–3962. doi: 10.1021/ie800275k

Zhang T. C., Surampalli R. Y., Lai K. C. K., Hu Z., Tyagi R. D. and Lo I. M. C. (2009). Nanotechnologies for Water Environment Applications. American Society of Civil Engineers (ASCE), Virginia, USA, doi: 10.1061/9780784410301.

Zhang Y., Yang M., Dou X.-M., He H. and Wang D.-S. (2005). Arsenate adsorption on an Fe – Ce bimetal oxide adsorbent: role of surface properties. *Environmental Science and Technology*, **39**(18), 7246–7253. doi: 10.1021/es050775d

Zhong L. S., Hu J. S., Liang H. P., Cao A. M., Song W. G. and Wan L. J. (2006). Self-assembled 3D flowerlike iron oxide nanostructures and their application in water treatment. *Advanced Materials*, **18**(18), 2426–2431. doi: 10.1002/adma.200600504

Chapter 4

Hierarchical materials as a design concept for multifunctional membranes

Christopher A. Crock, Brian J. Starr, and Volodymyr V. Tarabara

4.1 INTRODUCTION

We are surrounded by hierarchies that manifest themselves in many ways. Examples abound in both natural and social systems from ethics to government to computer code to biology. Maslow's hierarchy of needs represents an ordered set such that the most pressing need must be met before the next highest need. Democratic systems are organized in a way that all citizens can participate in government by electing representatives at multiple levels from local municipalities to state legislatures to federal legislatures. Computer science employs the principle of hierarchy in object-oriented programming wherein simple objects or data types are elements of more complex data structures. Biology offers a beautiful example of a hierarchy – a DNA molecule – where the complexity order is descended level-by-level from the double helix of biopolymers to nucleotides to nucleic bases to the atoms of which they are made.

"There is plenty of room at the bottom", speaking to the ability to manipulate matter at such small scales, Richard Feynman famously proclaimed in 1959 and pointed in the direction where the nanotechnology has been heading since mid-1990s. Indeed, the advantage of such compositional hierarchies may be especially fruitful in nanotechnology that allows us to manipulate matter at finer scales and lower levels of dimensional hierarchy. This provides the ability to assign to each level in a complex system an autonomous functionality.

In 1963, Loeb and Sourirajan developed the integrally skinned asymmetric membrane (Loeb & Sourirajan, 1963), a breakthrough in membrane technology that ushered in the commercial use of membranes for municipal water treatment.

Asymmetry arises from the ordered hierarchy of pores, where the size of pores in the membrane "skin" are several orders of magnitude smaller than pores in the support layer (Figure 4.1). The asymmetric design reduces the hydraulic resistance of the membrane while maintaining its rejection capability. Recently, significant advances have been made in the design of nanocomposite membranes that show great promise for drinking water and wastewater treatment by alleviating disadvantages of membranes (e.g. membrane fouling) and enabling efficient membrane-based reactions.

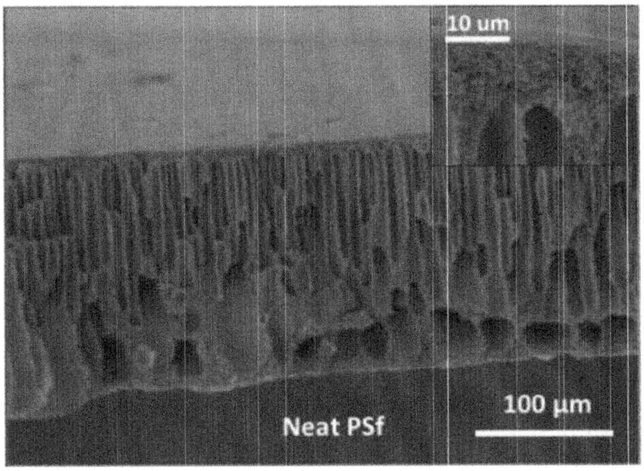

Figure 4.1 Cross-section of an asymmetric polysulfone membrane fabricated via phase inversion.

This chapter briefly overviews three emerging nanocomposite membrane technologies where hierarchical design of membranes is employed: 1) membranes coated with photocatalytic nanomaterials for hybrid disinfection – membrane filtration, 2) nanocomposite membranes with embedded hierarchical nanocatalysts as catalytic membrane reactors, and 3) superhydrophobic membranes for membrane distillation (MD). The examples illustrate how the hierarchical design of membrane materials can lead to improved separation properties and provide additional degrees of freedom for performance optimization.

4.2 PHOTOCATALYTIC MEMBRANES AND MEMBRANE REACTORS

Photocatalytic membrane reactor (PMR) technology combines TiO_2 membranes and UV light to enable concurrent separation and photocatalytic oxidation (Bai *et al.* 2012; Choi *et al.* 2006; Zhang *et al.* 2008, 2009). Most of the PMR research work has been on treating chemical contaminants with only several reports available on the application of PMR to pathogen inactivation (Choi *et al.* 2007; Zhang *et al.* 2009; Ma *et al.* 2009; Liu *et al.* 2012; Goei & Lim, 2014; Guo *et al.* 2015). The added photocatalytic function was also shown to improve biofouling resistance of membranes (e.g. (Damodar *et al.* 2009)). In the first application of photocatalytic membranes for virus removal and inactivation. Guo *et al.* (2015) showed the hybrid photocatalytic microfiltration–UV process was considerably more effective in inactivating the virus than microfiltration and UV disinfection applied sequentially (Figure 4.2).

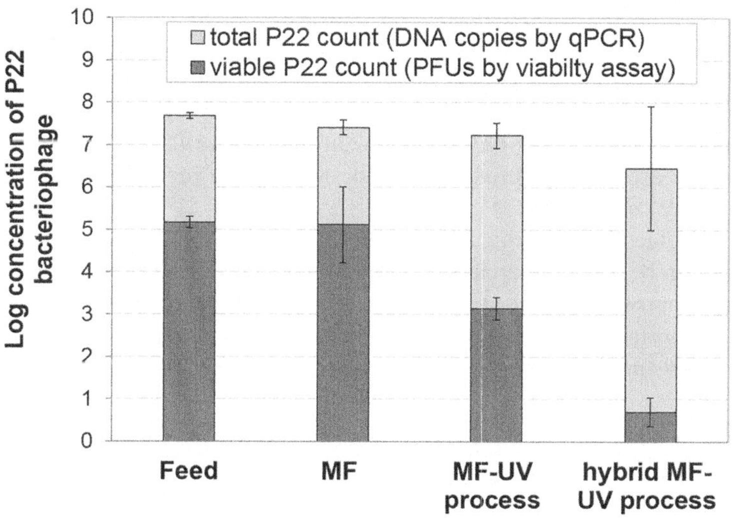

Figure 4.2 Log concentration of viable and total bacteriophage P22 in the feed solution and in the permeate 30 min into the filtration process. Error bars correspond to standard deviations (Guo *et al.* 2015).

Photocatalytic oxidation results when UV light strikes the surface of the catalyst, commonly TiO_2, resulting in excited electrons and remaining electron–holes. Hydroxyl free radicals formed by holes reacting with water can degrade organic micropollutants and inactivate pathogens complementing the effect of direct UV light. Compared to its bulk counterpart, nanoscale TiO_2 is superior in the production of OH* radicals (i.e. the photocatalytic oxidation efficiency is higher) due to a larger surface area and the quantum size effect. Consequently, significant efforts have been devoted to the fabrication and application of TiO_2 nanocomposite membranes.

Zhang *et al.* fabricated freestanding photocatalytic TiO_2 nanowire membranes by exploiting dimensionally different TiO_2 nanowires where separation performance and permeability were determined by the smaller nanowire dimension (Zhang *et al.*

2009). Hierarchical porosity, due to the use of nanowires of different diameters, presented additional degrees of freedom for controlling the membrane selectivity, where smaller (10 nm) diameter nanowires provided an excellent separation layer atop the larger (20 nm to 100 nm) diameter nanowires.

A challenge in photocatalysis is the recombination of charge carriers (i.e. electron–hole pairs) responsible for redox reactions. Liu and researchers designed a multi-component photocatalyst using TiO_2 and SnO_2, to prevent or slow down the recombination of electron–hole pairs, where SnO_2 and TiO_2 accumulate photogenerated electrons and holes, respectively (Liu *et al.* 2007). For the overall reactivity to be significantly enhanced, there must be access to both the accumulated electrons and holes, therefore, both TiO_2 and SnO_2 surfaces must be freely available (Hoffmann *et al.* 1995). Similarly, Bai *et al.* fabricated photocatalytic membranes that were fouling resistant and antibacterial using ZnO nanorods grown from TiO_2 nanowires (Bai *et al.* 2012). The additional hierarchy level, ZnO nanorods, improved the photocatalytic performance by increasing the total surface area and reactive sites, and slowing the recombination of charge carriers (Bai *et al.* 2012).

By hierarchically designing photocatalysts and photocatalytic membranes their efficiency can be greatly increased. The hierarchical design can promote UV light scattering, improve separation properties, and minimize the recombination of charge carriers. Importantly, by assigning the separation and photocatalytic functions to different levels in the hierarchy, each function can be optimized separately.

4.3 HIERARCHICALLY DESIGNED NANOCATALYSTS FOR CATALYTIC MEMBRANES

Small neutral molecules and some important ions such as nitrate, nitrite, and As (V) are rejected relatively poorly even by dense membranes in the nanofiltration and reverse osmosis range. Consequently, current conventional membrane treatment processes are poor candidates for the removal of these chemicals. By introducing nanocatalysts, however, and in the presence of reducing agents, membranes can be used as flow-through catalytic reactors that efficiently remove these chemicals with concurrent separation.

Crock *et al.* developed catalytic UF membranes by embedding hierarchical nanostructures – nanogold deposited on exfoliated graphite nanoplatelets (xGnPs) – in a polymeric membrane, reducing nitrophenol to aminophenol (Crock *et al.* 2013). The hierarchical nature of these nanocomposite membranes allowed for membrane optimization, such that xGnPs largely determined the membrane's separation performance and permeability, and nanogold provided catalytic functionality (Crock *et al.* 2013). The same reaction of catalytic reduction of nitrophenol was performed by Wang *et al.*, who used a hierarchically structured membrane wherein gold nanoparticles were attached to vertically aligned carbon nanotube arrays grown on a stainless steel mesh (Wang *et al.* 2013).

Table 4.1 Examples of hierarchical catalysts for catalytic reactions.

Catalytic Reaction	Reactor Type	Catalyst Type	Hierarchy Level 1	Hierarchy Level 2	Hierarchy Level 3	Reaction Rate Constant, k_{cat}	Nanocatalyst Design	Reference
4-nitrophenol to 4-aminophenol	Flow-through	Embedded in Al_2O_3 membrane	Au NPs	–	–	18 μm/s		
		xGnP supported; Embedded in polysulfone membrane	xGnPs	Au NPs	–	2–12 μm/s		(Crock *et al.* 2013)
TCE dechlorination	Batch	Unsupported	Au NPs	Pd islands	–	1956 $min^{-1} (g_{Pd}/L)^{-1}$		(Nutt *et al.* 2005; Nutt *et al.* 2006)
		MgO supported	MgO	Au NPs	Pd islands	1670 $min^{-1} (g_{Pd}/L)^{-1}$		
		SiO_2 supported	SiO_2			983 $min^{-1} (g_{Pd}/L)^{-1}$		
		Al_2O_3 supported	Al_2O_3	Pd NPs	–	47 $min^{-1} (g_{Pd}/L)^{-1}$		
TCE dechlorination	Flow-through	xGnP supported; Embedded in polysulfone membrane	xGnPs	Au NPs	Pd islands	559 $min^{-1} (g_{Pd}/L)^{-1}$ 18255 $(M_{H2} \cdot s)^{-1} (g_{Pd}/L)^{-1}$	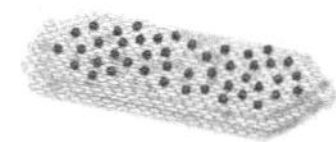	(Crock & Tarabara, 2016)
NO_3^- *denitrification*	Batch	Al_2O_3 supported	Al_2O_3	Cu NPs	Pd islands	3.63 $min^{-1} (g_{Me}/L)^{-1}$		(Chaplin *et al.* 2006)
		C supported	C	–	–	5.12 $min^{-1} (g_{Me}/L)^{-1}$		(Lemaignen *et al.* 2002)
		PVP supported	PVP	–	–	0.061 $min^{-1} (g_{Me}/L)^{-1}$		(Guy *et al.* 2009)
		C supported	C	Sn NPs	Pd islands	6.08 $min^{-1} (g_{Me}/L)^{-1}$		(Gavagnin *et al.* 2009)
NO_3^- and NO_2^- denitrification (proposed)	Flow-through	Graphene supported; Embedded in a membrane	Graphene	Au NPs	Cu islands	–	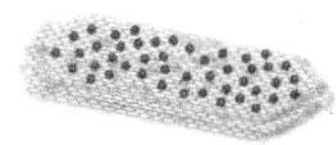	–

Nutt and researchers designed bimetallic Pd-on-Au nanocatalysts for trichloroethylene dechlorination that were fabricated in a hierarchical fashion (Nutt *et al.* 2005), where Pd (hierarchy level 1) was the catalyst and Au (hierarchy level 2) was the promoter metal (Table 4.1). It was later shown that the degree to which Pd covered Au regulated the efficiency of catalytic TCE reduction. Additionally, Au completely prevented chlorine poisoning and partially prevented sulfide poisoning of the Pd catalyst (Nutt *et al.* 2006). Compared to Pd catalysts supported on Al_2O_3, the Pd-on-Au catalysts increased the first order reaction rate constant from 47 to 1956 $L/g_{Pd}/min$ (Nutt *et al.* 2006).

While tests were performed in batch reactors, application of these catalysts in a membrane reactor is promising because the high reactivity should help to negotiate the conflicting demands of throughput and sufficient detection time to complete the reaction. Crock and Tarabara designed xGnP-supported metallic (Pd) and bimetallic (Pd-Au) catalysts that enabled membrane-based TCE hydrodechlorination with rate constants up to 81 times higher than those obtained for the commercial Pd/Al_2O_3 catalyst; the resulting nanocomposite membranes removed 96% of TCE at the permeate flux of 47.4 $L/(m^2{\cdot}h{\cdot}bar)$ (Crock & Tarabara, 2016).

Catalytic denitrification has been explored using Pd catalysts, but for nitrate to be reduced to nitrite a bimetallic catalyst consisting of a promoter metal and the Pd catalysts must be used (Horold *et al.* 1993). The promoter metal can be thought of as level 2 in the hierarchical design of the bimetallic catalyst for denitrification. Catalysts for the denitrification of nitrate and nitrite were fabricated using supported Pd-Cu nanoparticles on PVP colloids (Guy *et al.* 2009). The Pd-Cu/PVP nanocatalyst can be viewed as consisting of three hierarchy levels: 1) the PVP support, 2) the Pd catalysts, and 3) the Cu promoter metal for the reduction of nitrate into nitrite. The relative compositions of Pd and Cu on the PVP support were manipulated, and it was determined that the highest performing catalysts was a mixture of 70% to 80% Pd and 20% to 30% Cu (Guy *et al.* 2009). By choosing different supports for Pd-Cu catalysts used in batch reactions, the reaction rate constants were in (0.061 to 5.12) $L/min/g_{metal}$ range (Guy *et al.* 2009; Chaplin *et al.* 2006; Gavagnin *et al.* 2002; Lemaignen *et al.* 2002).

Exploring the design of nanocatalysts can lead to an improvement in the efficiency of the catalytic reactions. By embedding these hierarchically designed catalysts in membranes, catalysis for water treatment can be made more appealing by 1) decoupling membrane separation and reaction functions, 2) obviating the need to disperse and subsequently remove catalysts from batch reactors, and 3) improving diffusion limitations of the mass transfer of pollutants to catalysts. Moreover, membrane reactors eliminate the need to dispose of concentrate waste by transforming pollutants to nontoxic compounds.

4.4 SUPERHYDROPHOBIC MEMBRANES

Superhydrophobicity results from the combination of appropriate surface roughness features and low surface energy, which can be described by the Wenzel (Wenzel, 1936) and Cassie-Baxter models (Cassie & Baxter, 1944). Mimicking natural surfaces such as that of the lotus leaf (Barthlott & Ehler, 1977), superhydrophobic surfaces have self-cleaning characteristics and may be employed to design fouling-resistant membranes. Due to the low contact area between a water droplet and the superhydrophobic surface, water droplets roll off the surface while picking up surface foulants. The self-cleaning property can be viewed as an "additional function", and is a strong motivation for using superhydrophobic membranes in a membrane distillation (MD) process where low surface energy and fouling-resistant membranes are required (Table 4.2).

Significant advantages of MD over other desalination technologies are high rejection and low operating temperatures. Waste heat from industrial activities or solar energy can be used as an energy source to bring the feed water to temperatures appropriate for MD, thereby decreasing energy inputs. An ideal MD membrane has straight pores with large pore diameters contributing to high flux, while also having low thermal conductivity and high fouling resistance (Ma *et al.* 2009). Superhydrophobic membranes can enhance the MD process by increasing the liquid entry pressure, thus allowing for larger pore diameters, and mitigating membrane fouling. Currently, MD technology is mostly in the lab-scale phase, but there has been some work at the pilot scale (Guillen-Burrieza *et al.* 2012; Sirkar *et al.* 2009; Walton *et al.* 2004).

Hong *et al.* made use of hierarchical roughness features afforded by block copolymer micelles grafted to colloidal silica films making them superhydrophobic where the contact angle can be tuned from 122 to 171° as a function of nanoscale surface features on silica colloids (Hong *et al.* 2007). Similarly, Razmjou and researchers created hierarchical surface roughness features by coating PVDF membranes for MD with TiO_2 nanoparticles, which then were fluorinated with perfluorododecyltrichlorosilane (FTCS) (Razmjou *et al.* 2012). The contact angle (CA) of these neat PVDF membranes (CA = 125°) decreased to 98° with the deposition of TiO_2 particles that created microscale roughness features, but the PVDF membrane became superhydrophobic (CA = 163°) after fluorinating TiO_2 with FTCS to create nanoscale roughness features on the surface of TiO_2 (Razmjou *et al.* 2012). Zheng and researchers used hierarchical roughness features afforded by nanoscale $(CH_3)_2SiCl_2/CH_3SiCl_3$ filaments on microscale PVDF aggregates to endow PVDF membranes with superhydrophobicity and self-cleaning properties (Zheng *et al.* 2011). Ma and researchers designed superhydrophobic glass membranes for MD that resisted fouling by etching hollow-cylindrical and spiked arrays onto a glass surface (Ma *et al.* 2009) (Table 4.2).

Table 4.2 Superhydrophobic surfaces designed with surface roughness hierarchy.

Morphological Features		Contact angle Hysteresis $(\theta_{adv} - \theta_{rec})$,°	Contact Angle,°	Surface Morphology	Reference
Microscale	**Nanoscale**				
Etched fluorocarbon coated SiO_2, 4.1 μm	None	8 to 28	156 to 165		(Oner & McCarthy, 2000)
Etched fluorocarbon coated SiO_2, 9.5 μm	None	4 to 12	156 to 165		
CNT arrays, ~4.5 μm	CNT array surface roughness	0 to 40	154 to 165	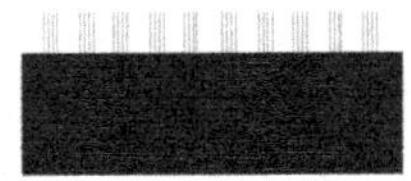	
CNT arrays, ~9.8 μm	CNT array surface roughness	0 to 1	154 to 165		
Etched fluorocarbon coated SiO_2, ~4.1 μm	CNT array surface roughness	0 to 1	154 to 165	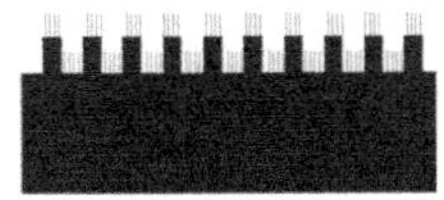	
Etched fluorocarbon coated SiO_2, ~9.5 μm	CNT array surface roughness	0 to 1	154 to 165		
Fluorine coated glass micro-channels	Fluorine coated glass nano-spikes	Not reported	135 to 165		(Ma *et al.* 2009)
TiO_2 coated PVDF membrane	None	47	98	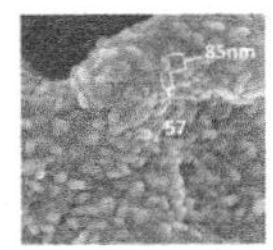	(Razmjou *et al.* 2012)
TiO_2 coated PVDF membrane	Fluorinated TiO_2 coated PVDF membrane	2	163	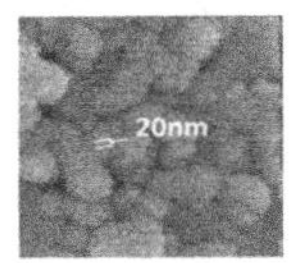	

The importance of the hierarchical structure can be described by the decrease in the interface between liquid and solid (f_S), as shown by the Cassie-Baxter model where contact angles approach 180° as f_S goes to 0. Roughening the surface at the microscale decreases f_S by introducing air gaps between the liquid-solid interfaces, but at this scale it is difficult to obtain a stable superhydrophobic surface (Hong *et al.* 2007) with minimal hysteresis. Without nanoscale roughness, a receding drop on microscale roughness features can leave a layer of water behind, increasing the solid interface area (Razmjou *et al.* 2012). Öner and researchers showed that the maximum scale for improving hydrophobicity for square roughness features was near 32 μm (Oner & McCarthy, 2000). After introducing nanoscale roughness to the microscale features, a stable superhydrophobic surface can be achieved decreasing the contact angle hysteresis, the difference between receding and advancing contact angles, to less than 1° (Razmjou *et al.* 2012; Zhu *et al.* 2005; Oner & McCarthy, 2000).

By applying compositional hierarchy to the design of MD membranes, the combination of microscale and nanoscale roughness features (e.g. (Meng *et al.* 2014; Tijing *et al.* 2016)) can be adjusted at each hierarchy level (i.e. the microscale and nanoscale) to impart superhydrophobicity, which can 1) increase membrane permeability due to higher LEPs, 2) mitigate fouling due to decreased water-membrane contact area, and 3) enable self-cleaning due to low contact angle hysteresis. These characteristics of MD membranes improve the likelihood of MD becoming a practical water treatment technology.

4.5 FUTURE RESEARCH

Future research on nanocomposite membranes for water treatment will be focused at the interface of materials science and membrane science. By designing nanocomposites using compositional hierarchy, the efficiency of membrane functionality and the degrees of freedom in membrane fabrication can be increased. Photocatalytic membrane reactors may benefit from multiscale design by optimizing its catalytic and separation functions independently as well by enabling finer control over the geometry of the reaction zone. In membrane reactors for reductive catalysis, incorporating additional levels in the hierarchy and optimizing the combination of the catalyst support, catalyst, and promoter metal can help achieve better catalytic performance. In membrane distillation, multi-level surface roughness is critical for designing a superhydrophobic membrane surface that resists fouling and promotes self-cleaning.

4.6 ACKNOWLEDGEMENTS

This material is based upon work supported in part by the Paul L. Busch award from the Water Environment Research Foundation and the NSF Partnerships for International Education and Research grant IIA-1243433.

4.7 REFERENCES

Bai H. W., Liu, Z. Y. and Sun D. D. (2012). A hierarchically structured and multifunctional membrane for water treatment. *Applied Catalysis B: Environmental*, **111**, 571–577.

Cassie A. B. D. and Baxter S. (1944). Wettability of porous surfaces. *Transactions of the Faraday Society*, **40**, 0546–0550.

Chaplin B. P., Roundy E., Guy K. A., Shapley J. R. and Werth C. J. (2006). Effects of natural water ions and humic acid on catalytic nitrate reduction kinetics using an alumina supported Pd-Cu catalyst. *Environmental Science and Technology*, **40**, 3075–3081.

Choi H., Sofranko A. C. and Dionysiou D. D. (2006). Nanocrystalline TiO2 photocatalytic membranes with a hierarchical mesoporous multilayer structure: Synthesis, characterization, and multifunction. *Advanced Functional Materials*, **16**, 1067–1074.

Choi H., Antoniou M. G., de la Cruz A. A., Stathatos E. and Dionysiou D. D. (2007). Photocatalytic TiO2 films and membranes for the development of efficient wastewater treatment and reuse systems. *Desalination*, **202**, 199–206.

Crock C. A. and Tarabara V. V. (2016). Pd and Pd-Au nanocatalysts supported on exfoliated graphite for high throughput dehalogenation by nanocomposite membranes. *Environmental Science: Nano*, **3**, 453–461.

Crock C. A., Rogensues A. R., Shan W. Q. and Tarabara V. V. (2013). Polymer nanocomposites with graphene-based hierarchical fillers as materials for multifunctional water treatment membranes. *Water Research*, **47**, 3984–3996.

Damodar R. A., You S.-J. and Chou H.-H. (2009). Study the self cleaning, antibacterial and photocatalytic properties of TiO2 entrapped PVDF membranes. *Journal of Hazardous Materials*, **172**, 1321–1328.

Gavagnin R., Biasetto L., Pinna F. and Strukul G. (2002). Nitrate removal in drinking waters: the effect of tin oxides in the catalytic hydrogenation of nitrate by Pd/SnO2 catalysts. *Applied Catalysis B: Environmental*, **38**, 91–99.

Goei R. and Lim T.-T. (2014). Ag-decorated TiO2 photocatalytic membrane with hierarchical architecture: photocatalytic and antibacterial activities. *Water Research*, **59**, 207–218.

Guillen-Burrieza E., Zaragoza G., Miralles-Cuevas S. and Blanco J. (2012). Experimental evaluation of two pilot-scale membrane distillation modules used for solar desalination. *Journal of Membrane Science*, **409**, 264–275.

Guo B., Pasco E. V. and Xagoraraki I. and Tarabara V. V. (2015). Virus removal and inactivation in a hybrid microfiltration-UV process with a photocatalytic membrane, Separ. *Purification Technology*, **149**, 245–254.

Guy K. A., Xu H. P., Yang J. C., Werth C. J. and Shapley J. R. (2009). Catalytic nitrate and nitrite reduction with Pd-Cu/PVP colloids in water: Composition, structure, and reactivity correlations. *The Journal of Physical Chemistry C*, **113**, 8177–8185.

Hoffmann M. R., Martin S. T., Choi W. and Bahnemann D. W. (1995). Environmental applications of semiconductor photocatalysis. *Chemical Reviews*, **95**, 69–96.

Hong J., Bae W. K., Lee H., Oh S., Char K., Caruso F. and Cho J. (2007). Tunable superhydrophobic and optical properties of colloidal films coated with block-copolymer-micelles/micelle-multilayers. *Advanced Materials*, **19**, 4364-+.

Horold S., Tacke T. and Vorlop K. D. (1993). Catalytical removal of nitrate and nitrite from drinking-water 1. Screening for hydrogenation catalysts and influence of reaction conditions on activity and selectivity. *Environmental Technology*, **14**, 931–939.

Lemaignen L., Tong C., Begon V., Burch R. and Chadwick D. (2002). Catalytic denitrification of water with palladium-based catalysts supported on activated carbons. *Catalysis Today*, **75**, 43–48.

Liu L., Liu Z. Y., Bai H. W. and Sun D. D. (2012). Concurrent filtration and solar photocatalytic disinfection/degradation using high-performance Ag/TiO2 nanofiber membrane. *Water Research*, **46**, 1101–1112.

Liu Z. Y., Sun D. D. L., Guo P. and Leckie J. O. (2007). An efficient bicomponent TiO2/SnO2 nanofiber photocatalyst fabricated by electrospinning with a side-by-side dual spinneret method. *Nano Letters*, **7**, 1081–1085.

Loeb S. and Sourirajan S. (1963). Sea water demineralization by means of an osmotic membrane. *Advances in Chemistry*, **38**, 117–132.

Ma N., Fan X., Quan X. and Zhang Y. (2009). Ag-TiO2/HAP/Al2O3 bioceramic composite membrane: fabrication, characterization and bactericidal activity. *Journal of Membrane Science*, **336**, 109–117.

Ma Z. Y., Hong Y., Ma L. Y. and Su M. (2009). Superhydrophobic membranes with ordered arrays of nanospiked microchannels for water desalination. *Langmuir*, **25**, 5446–5450.

Barthlott W. and Ehler N. (1977). Raster-Elektronenmikroskopie der Epidermis-Oberflächen von Spermatophyten. *Tropische und subtropische Pflanzenwelt (Akad. Wiss. Lit. Mainz)*, **19**, 110.

Meng S., Mansouri J., Yea Y. and Chen V. (2014). Effect of templating agents on the properties and membrane distillation performance of TiO2-coated PVDF membranes. *Journal of Membrane Science*, **450**, 48–59.

Nutt M. O., Heck K. N., Alvarez P. and Wong M. S. (2006). Improved Pd-on-Au bimetallic nanoparticle catalysts for aqueous-phase trichloroethene hydrodechlorination. *Applied Catalysis B: Environmental*, **69**, 115–125.

Nutt M. O., Hughes J. B. and Wong M. S. (2005). Designing Pd-on-Au bimetallic nanoparticle catalysts for trichloroethene hydrodechlorination. *Environmental Science and Technology*, **39**, 1346–1353.

Oner D. and McCarthy T. J. (2000). Ultrahydrophobic surfaces. Effects of topography length scales on wettability. *Langmuir*, **16**, 7777–7782.

Razmjou A., Arifin E., Dong G. X., Mansouri J. and Chen V. (2012). Superhydrophobic modification of TiO2 nanocomposite PVDF membranes for applications in membrane distillation. *Journal of Membrane Science*, **415**, 850–863.

Sirkar K. and Song L. 2009. Pilot-scale studies for direct contact membrane distillation-based desalination process, in, New Jersey Institute of Technology.

Tijing L. D., Woo Y. C., Shim W.-G., He T., Choi J.-S., Kim S.-H. and Shon H. K. (2016). Superhydrophobic nanofiber membrane containing carbon nanotubes for high-performance direct contact membrane distillation. *Journal of Membrane Science*, **502**, 158–170.

Walton J., Lu H., Turner C., Solis S. and Hein H. (2004). Solar and waste heat desalination by membrane distillation, in, University of Texas at El Paso.

Wang H., Dong Z. and Na C. (2013). Hierarchical carbon nanotube membrane-supported gold nanoparticles for rapid catalytic reduction of p-nitrophenol. *ACS Sustainable Chemistry & Engineering*, **1**, 746–752.

Wenzel R. N. (1936). Resistance of solid surfaces to wetting by water, Ind. *Engineering Chemistry*, **28**, 988–994.

Zhang H., Zhao H., Liu P., Zhang S. and Li G. (2009). Direct growth of hierarchically structured titanate nanotube filtration membrane for removal of waterborne pathogens. Journal of Membrane Science, **343**, 212–218.

Zhang X. W., Du A. J., Lee P. F., Sun D. D. and Leckie J. O. (2008). TiO2 nanowire membrane for concurrent filtration and photocatalytic oxidation of humic acid in water. *Journal of Membrane Science*, **313**, 44–51.

Zhang, X. W., Zhang, T., Ng, J. and Sun, D. D. (2009). High-performance multifunctional TiO2 nanowire ultrafiltration membrane with a hierarchical layer structure for water treatment. *Advanced Functional Materials*, **19**, 3731–3736.

Zheng Z. R., Gu Z. Y., Huo R. T. and Luo Z. S. (2011). Fabrication of self-cleaning poly(vinylidene fluoride) membrane with micro/nanoscaled two-tier roughness. *Journal of Applied Polymer Science*, **122**, 1268–1274.

Zhu L. B., Xiu Y. H., Xu J. W., Tamirisa P. A., Hess D. W. and Wong C. P. (2005). Superhydrophobicity on two-tier rough surfaces fabricated by controlled growth of aligned carbon nanotube arrays coated with fluorocarbon. *Langmuir*, **21**, 11208–11212.

Chapter 5

Smart membrane materials for controllable oil-water separation

Lianbin Zhang and Peng Wang

5.1 INTRODUCTION

Oil is one of the most important energy sources in the world and the oil-related industries play a critical role in the development of modern industry and economy. However, the production, transportation, and utilization of the oil-related products all pose potential environmental concerns due to the risk of the spillage into the environment (Short *et al.* 2003; Dubansky *et al.* 2013). For instance, oily wastewater, resulting from various industrial processes such as metallurgy, food, textile, leather, and petrochemical, has become a common pollution source all over the world (Kajitvichyanukul *et al.* 2006; Chan *et al.* 2009; Asatekin *et al.* 2009). On the other hand, frequent oil-spill incidents that release a large amount of liquid oils into the marine environments bring severe risks to marine ecosystem and to human society as well (McNutt *et al.* 2012). To deal with oil-polluted waters, a number of strategies have been widely adopted. Direct burning of the oils has been used to treat oil spill and oily waters as an initial response means when the oil slick is thick. However, this method generates fire hazards and also brings in undesired air pollution problem. Different from the direct burning of the oil, separation of oil and water has been gaining popularity as a more effective strategy because it allows for collection and even reuse of the spilled oil (Kajitvichyanukul *et al.* 2006). To this end, several methods have been in practice or proposed, including mechanical separation (e.g., the use of skimmers, booms, pumps, mechanical separators, etc.), wettability-based separation (e.g., the use of hydrophobic/oleophilic materials to separation oil from water), chemical dispersant-based approach, microorganism-based selective digestion of oil, etc (Nordvik *et al.* 1996; El-Kayar *et al.* 1993; Mostefa *et al.* 2004; Head *et al.* 2006; Bai *et al.* 2011; Xue *et al.* 2014). However, most of the above-mentioned separation methods often require tedious operations, are energy-intensive, or time-consuming. Furthermore, these methods rarely achieve complete separation of oil from water due to their intrinsic limitations.

On the other hand, membrane–based separation, a separation technology widely used in industry, is advantageous in oil-water separation due to its easy scale-up, mild operating conditions, absence of additives, and lower energy consumption (Drioli *et al.* 2010). The mechanism behind the membrane separation technology is simple, where a membrane works as a selective barrier and permits the separation of certain species in a fluid by combination of sieving and sorption diffusion mechanism. Separation is achieved by selectively passing (permeating) one or more components of a stream through the membrane while retarding the passage of one or more other components. Membrane separation has been widely adopted in water purification and fresh water production. Recently, with the development in colloidal and interface sciences, many bio-inspired functional membrane materials with superwetting capability, including superhydrophilic-underwater superoleophobic, superhydrophobic-superoleophilic, superhydrophobic-superoleophobic, etc., have been developed and these superwetting capability of the materials is the result of a combination of unique micro-nano hierarchical surface rough structure and surface chemistry (Sun *et al.* 2005; Sahoo *et al.* 2014; Zhu *et al.* 2014; Chu *et al.* 2015; Wang *et al.* 2015). Due to their unique surface wetting behaviors, these superwetting membrane materials can selectively attract oil (or water) while completely repelling water (or oil), and have thus shown tremendous potentials in the application of oil-water separation as they permit gravity-driven oil-water separation. For instance, inspired by the superhydrophobic lotus leaf, Jiang and coworkers (2004) first prepared superhydrophobic and superoleophilic mesh and applied it for gravity-driven oil-water separation. In this work, the mesh was prepared by depositing polytetrafluoroethylene (PTFE), a well-known low-surface-energy material, on a commercial stainless steel mesh and the coated mesh showed water contact angle greater than 150°, and diesel contact angle of ~0°. The drastically different water and oil wettability led to an effective separation of oil and water mixture by allowing oil phase to pass through while retaining water on top of it. Later, researchers gradually realized that although this superhydrophobic and superoleophilic material was effective in separating oil-water mixture, this 'oil-removing' type of materials could be easily fouled or even blocked by oils because of their intrinsic oleophilic property. Oils, especially high-viscosity oils, would adhere to the membranes and thus seriously affect the separation efficiency and shortens the lifetime of these membranes. In addition, the adhered or absorbed oils can be hard to remove from the membranes, leading to secondary pollution during the post-treatment process. To solve this problem, Jiang and coworkers (2011), inspired by oil-repellent capability of fish scale (Liu *et al.* 2009), fabricated a novel superhydrophilic and underwater superoleophobic hydrogel coated mesh (Figure 5.1). This 'water-removing' type of material has a completely opposite wettability to traditional hydrophobic and oleophilic ones, and thus overcomes the easy-fouling and hard-recycling limitations. In essence, (1) the 'water-removing' materials allow only water to pass and repel oil from making contact with the membranes, which effectively eliminates or reduces the possibility of membrane clogging by viscous oil; (2) they prevent the formation of a water barrier between the membranes and the oil phase, which would otherwise occur with the conventional hydrophobic and oleophilic separation (e.g., 'oil-removing') materials due to the fact that water is generally heavier than oil phase. This novel material has been showed to selectively separate water from various oil-water mixtures, such as vegetable oil, gasoline, diesel, and even crude oil/water mixtures with a high separation efficiency (>=99%) and without the need of extra power. Even since this work, a series of the hydrophilic and underwater superoleophobic membrane materials have been developed following similar scientific principle (Zhang *et al.* 2013; Zhu *et al.* 2013; Gao *et al.* 2014; He *et al.* 2015).

Figure 5.1 (a) Surface structures of fish scale. Inset in (a): shape of an oil droplet on fish scales in water, showing the superoleophobicity of the fish scales. Reprinted with permission from (Liu *et al.* Copyright 2009) John Wiley and Sons. (b) Oil/water separation of the hydrogel-coated mesh. The coated mesh was fixed between two glass tubes, the mixture of crude oil and water was put into the upper glass tube. Water selectively permeated through the coated mesh, while the oil was repelled and kept in the upper glass tube. Reprinted with permission from (Xue *et al.* Copyright 2011) John Wiley and Sons.

Considering the complexity of oil-spill incidents as well as the diversity of oil-water mixtures in practice, a controllable oil-water separation, in which either oil or water phase can be allowed to pass on demand, is highly desirable and clearly advantageous. In principle, such a controllable separation entails a smart surface with an adjustable oil (or water) wettability, or more desirably a smart surface that switches its oil wettability in response to external stimuli in aqueous media. To obtain a smart surface with switchable oil wettability, especially between superoleophobicity and superoleophilicity in aqueous media, the chemistry on the surface should be delicately designed such that it comprises both hydrophilic and oleophilic/hydrophobic characteristics, with either characteristic becoming dominantly exposed over the other in response to environmental conditions. The first demonstration of the smart surface was provided by Wang and coworkers in 2012 when they developed a smart surface with switchable superoleophilicity and superoleophobicity in aqueous media for controllable oil-water separation as illustrated in Figure 5.2 (Zhang *et al.* 2012). In their work, they achieved surfaces with switchable superoleophilicity and superoleophobicity in aqueous media by grafting onto inexpensive and easily available materials, including non-woven textiles and polyurethane sponges, a block copolymer comprising pH-responsive poly(2-vinylpyridine) and oleophilic/hydrophobic polydimethylsiloxane blocks (i.e., P2VP-b-PDMS). The P2VP block on the grafted block copolymer can alter its wettability and its conformation via protonation and deprotonation in response to the pH of the aqueous media, which in turn provides controllable and switchable access of oil by the oleophilic PDMS block on the surface, thus realizing a smart surface with switchable oil wettability in aqueous media. With such a functionalized surface, highly controllable oil-water separation was realized.

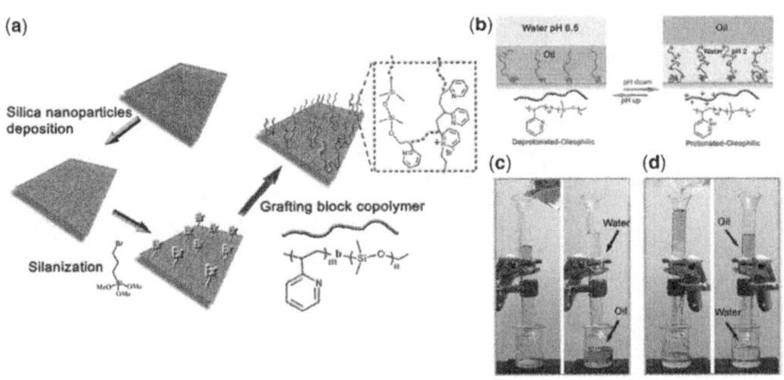

Figure 5.2 (a) Schematic showing the preparation strategy for a surface with switchable superoleophilicity and superoleophobicity on a non-woven textile. (b) Schematic diagrams for the switchable oil wettability of the P2VP-b-PDMS grafted textile. (c) Gasoline selectively passed through the textile, whereas the water remained in the upper glass tube when pH of water is 6.5. (d) If the functionalized textile was first wetted with acidic water (pH 2.0). Water selectively passed through the textile, whereas gasoline remained in the upper glass tube. Reprinted with permission from (Zhang *et al.* Copyright 2012) Nature Publishing Group.

Since then, a series of smart membranes with switchable wettability have been developed for the purpose of the controllable oil-water separation, and various kinds of external stimuli and treatments, including pH, light, temperature, electric field, solvent treatment, and gas treatment, have been employed to trigger the switch of the surface wettability and to realize the controllable oil-water separation. In this chapter, recent advances in the controllable or on-demand oil-water separation are presented. The theoretical principle to construct superwetting surfaces is briefly introduced in Section 5.2 first and a summary a perspective is presented at the end.

5.2 FUNDAMENTAL THEORY OF WETTABILITY OF SOLID MATERIALS

The wetting behavior of liquid on a flat solid surface is a function of the interfacial free energies among the solid/liquid (γ_{SL}), solid/vapour (γ_{SV}) and liquid/vapour (γ_{LV}) interfaces, which is commonly governed by the Young's equation (Equation 1) (Young, 1805):

$$\cos\theta = \frac{\gamma_{SV} - \gamma_{SL}}{\gamma_{LV}} \tag{1}$$

where, θ is the contact angle in the Young's mode (Figure 5.3a). Commonly, when the solid surface shows a liquid contact angle (CA) less than 90°, this solid material is defined as lyophilic, and when the solid surface shows a liquid CA more than 90°, this solid material is defined as lyophobic.

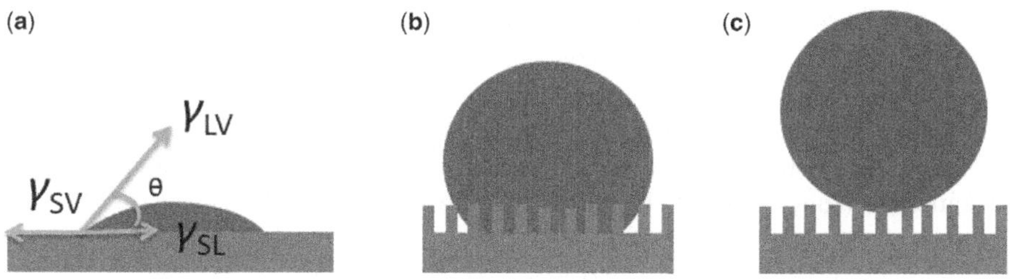

Figure 5.3 Schematic illustration of a droplet siting on different substrates. (a) on a flat substrate and (b, c) on rough substrates. Depending on the roughness of the surface, the droplet is either in the Wenzel regime (b) or the Cassie–Baxter regime (c).

The Young's equation is suited only for ideal flat surface, and in practice solid surface is usually not perfectly flat. Therefore, the effect of surface roughness has to be considered for evaluating surface wettability, and two different wetting regimes, the so-called Wenzel regime (Figure 5.3b) and the Cassie–Baxter regime

(Figure 5.3c), were developed to explain wetting behavior on a rough surface. In the Wenzel's regime (Wenzel, 1936), it is assumed that the liquid penetrates completely and contacts the interiors of the grooves on the solid surface, and the relation between the apparent (measured) angle θ^* and the ideal contact angle θ of a flat surface made of the same component is described by Equation (2), in which r is the ratio of the actual surface area to the projected area of the rough surface.

$$\cos\theta^* = r\cos\theta \tag{2}$$

In this Wenzel's model, surface roughness promotes either wettability ($\theta < 90°$) or nonwettability ($\theta > 90°$), depending on the chemical properties of the surface.

If air is trapped by the liquid to give a composite surface, wetting phenomenon can be described by the Cassie–Baxter model (Cassie *et al.* 1944) depicted by equation (3):

$$\cos\theta^* = f_1\cos\theta_1 + f_2\cos\theta_2 \tag{3}$$

where θ_1 and θ_2 are the contact angles of the liquid droplet upon the surface 1 (air) and surface 2, respectively. f_1 and f_2 are the apparent area fractions of surface 1 (air) and surface 2 in contact with the droplet ($f_1 + f_2 = 1$). Clearly, the corresponding θ^* is always higher than that of a flat surface made from the same material because the pores are filled with air. In this Cassie's regime, the liquid droplets stay in contact with only the highest asperities of the surface with a very limited contact area. Thus, the adhesion between the liquid droplet and the substrate surface is quite low. As a consequence, the liquid droplet can easily roll off the surface if the surface is slightly tilted relative to the horizontal plane.

Besides the solid/liquid/vapour system, when an oil droplet is deposited on a substrate surface that is under water, the Young equation can be modified to give Equation (4) (Jung *et al.* 2009; Liu *et al.* 2009):

$$\cos\theta_{ow} = \frac{\gamma_{OA}\cos\theta_O - \gamma_{WA}\cos\theta_W}{\gamma_{OW}} \tag{4}$$

in which γ_{OA}, γ_{WA}, and γ_{OW} are the oil/air, water/air, and oil/water interface tensions, respectively, and θ_O, θ_W, and θ_{OW} are the contact angles of oil in air, water in air, and oil in water, respectively. As predicted by Equation (4), for a hydrophilic surface ($\gamma_{SA} > \gamma_{SW}$), an oleophobic surface at the solid-water-oil interface can be obtained if $\gamma_{OA}\cos\theta_O$ is less than $\gamma_{WA}\cos\theta_W$. Since the surface tension of oil and organic liquids is much lower than that of water, most hydrophilic surfaces can be made oleophobic at the solid-water-oil interface. For a hydrophobic and an oleophilic surface at a solid-air-oil interface, an oleophilic surface at a solid-water-oil interface can be created. Similar to the situation in air, a flat underwater oleophobic surface may show underwater superoleophobic properties if an appropriate surface roughness is introduced. Therefore, to realize the controllable oil-water separation, suitable surface roughness as well as carefully tuned surface chemistry should be both taken in consideration when designing and fabricating the superwetting membranes.

5.3 CONTROLLABLE OIL-WATER SEPARATION WITH SUPERWETTING MEMBRANES

5.3.1 pH controlled oil-water separation

As discussed in the introduction section, Wang and co-workers for the first time reported a smart surface with switchable superoleophilicity and superoleophobicity in response to aqueous pHs (Figure 5.2). Recently, Shi and co-workers (2014) described the fabrication of a pH-responsive smart device for continuous *in situ* separation of oil/water/oil ternary mixtures without the need for *ex situ* treatments. In their work, a pH-responsive smart device was prepared on the copper foam by combining electroless deposition of silver aggregates on the copper foam surface with self-assembled monolayers (Figure 5.4a). The modified mixed thiols of $SH(CH_2)_9CH_3$ and $SH(CH_2)_{10}COOH$ at a molar ratio of 4:6 gave rise to the pH responsive wettability (Figure 5.4b). At such a ratio, the modified rough surface exhibited superhydrophobicity under acidic conditions because the carboxylic acid groups are mostly protonated, and thus the methyl groups play a dominant role in surface wettability; under alkaline condition, the majority of carboxylic acid groups is deprotonated and hydrated, contributing to a superhydrophilic surface property. Due to this pH-dependent wettability, the smart device was applied to oil/water separations both in air and underwater. Efficient separation and collection of the individual components of a complex oil/water/oil mixture was realized in a continuous process with no *ex situ* treatments as demonstrated in Figure 5.4c–4i. Similarly, Cheng *et al.* (2015) also used the mixed responsive thiol molecules to modify the $Cu(OH)_2$ nanorod-structured copper mesh substrate, and achieved the pH-triggered on-demand oil-water separation.

By using an electrospun poly(methyl methacrylate)-block-poly(4-vinylpyridine) (PMMA-*b*-P4VP) fiber membrane, Li *et al.* (2015) reported gravity-driven pH-controllable oil/water separations as shown in Figure 5.5. The block copolymer electrospun fiber membrane exhibited responsive oil wettability at different pHs as a consequence of the protonation and deprotonation of pyridyl groups. The other block, PMMA, is hydrophilic and underwater oleophilic. Oil selectively passed through the membrane, whereas water remained at the initial state; after the membrane was wetted with acidic water, a reverse separation was realized.

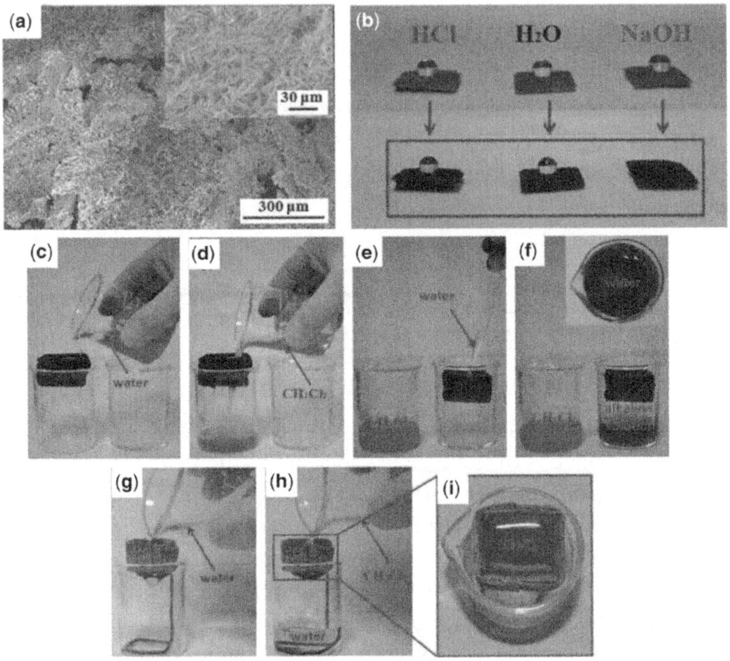

Figure 5.4 (a) Scanning electron micrograph of copper foam with modified silver aggregates. (b) Photographs display the original states of droplets of HCl, H_2O, and NaOH on the as-prepared smart surface (top images). After several minutes, the HCl and H_2O droplets are unchanged while the NaOH droplet spreads over the foam surface (bottom images). Photographs show the oil/water separation using the smart device (c, d). A mixture of water and green-dyed oil is poured into the smart device; the oil passes through into the supporting beaker. (e) Water is prevented from infiltrating the superhydrophobic surface and remains in the device box. (f) An aqueous alkaline solution, however, permeates through the device. (g–i) Water passes through the alkaline-treated device while oil is blocked. Reprinted with permission from (Ju *et al.* Copyright 2014) Nature Publishing Group.

5.3.2 Photo-controlled oil-water separation

As an alternative method, photo irradiation can also be used to tune the surface wettability so as to realize controllable oil-water separation. Tian *et al.* (2012) reported photo-induced water–oil separation based on a switchable superhydrophobicity–superhydrophilicity and underwater superoleophobicity on aligned ZnO nanorod array-coated mesh films, which showed an excellent controllability and high separation efficiency of different types of water–oil mixtures in an oil–water–solid three-phase system (Figure 5.6). The aligned ZnO nanorod array was prepared on the stainless steel mesh films using a two-step solution approach (Figure 5.6a). The micro/nanoscale hierarchical structured ZnO mesh films (Figure 5.6b–d) showed superhydrophobicity (after storage in the dark) and superhydrophilicity and underwater superoleophobicity (under illumination) (Figure 5.6e and 5.6f), and achieved the photo-induced water–oil separation as illustrated in Figure 5.6g and 5.6h. Such a photo responsive mesh membrane is promising in application in a micro-reaction system and controllable filtration.

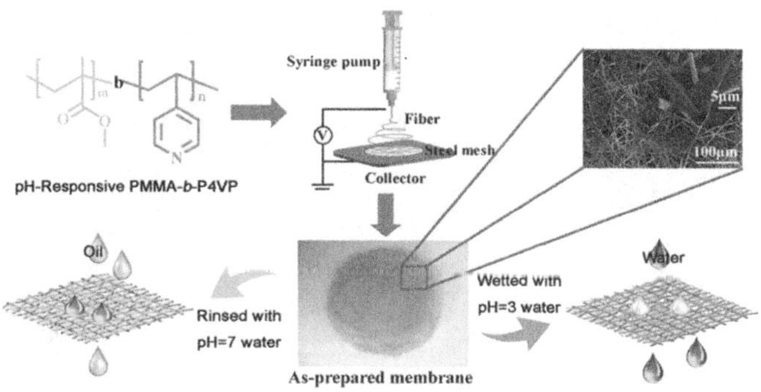

Figure 5.5 Schematic showing the fabrication of the block copolymer fiber membranes and its application for the controllable oil-water separation. Reprinted with permission from (Li *et al.* Copyright 2015) American Chemical Society.

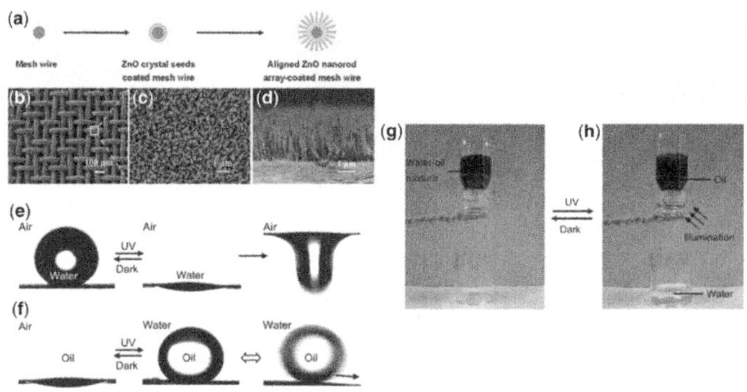

Figure 5.6 Schematic diagrams of the preparation process (a) and SEM images (b–d) of the as-prepared aligned ZnO nanorod array-coated stainless steel mesh films. Photo-induced surface wettability of the aligned ZnO nanorod arrays-coated mesh films. (e) Photographs of a water droplet on the coated mesh film after dark storage (left) and under UV irradiation (middle) in air with a contact angle of 155° and 0°, respectively. When more water is added, the water droplet will drop (right). (f) Photographs of an oil droplet (1,2-dichloroethane) on the mesh film in air (left) and underwater (middle) with a contact angle of 0° and 156°, respectively, and sliding angle of 2.6° (right). (g) and (h) Photo-induced water–oil separation. Reproduced with permission from (Tian *et al.* Copyright 2012) The Royal Society of Chemistry.

Recently, Kim *et al.* (2015) fabricated a UV-responsive nano-sponge consisting of hydrophobic hydrocarbon and hydrophilic TiO_2 nanoparticles for oil absorption or desorption. The hydrocarbon in the nano-sponge could selectively absorb oil in water, whereas the absorbed oil can be released back into the water by TiO_2 in response to UV irradiation as shown in Figure 5.7. In one experiment, the nano-sponge functionalized porous polydimethylsiloxane released more than 98% of the absorbed crude oil with UV irradiation and air-bubbling. It could be continuously reused without sacrificing its adsorption capacity.

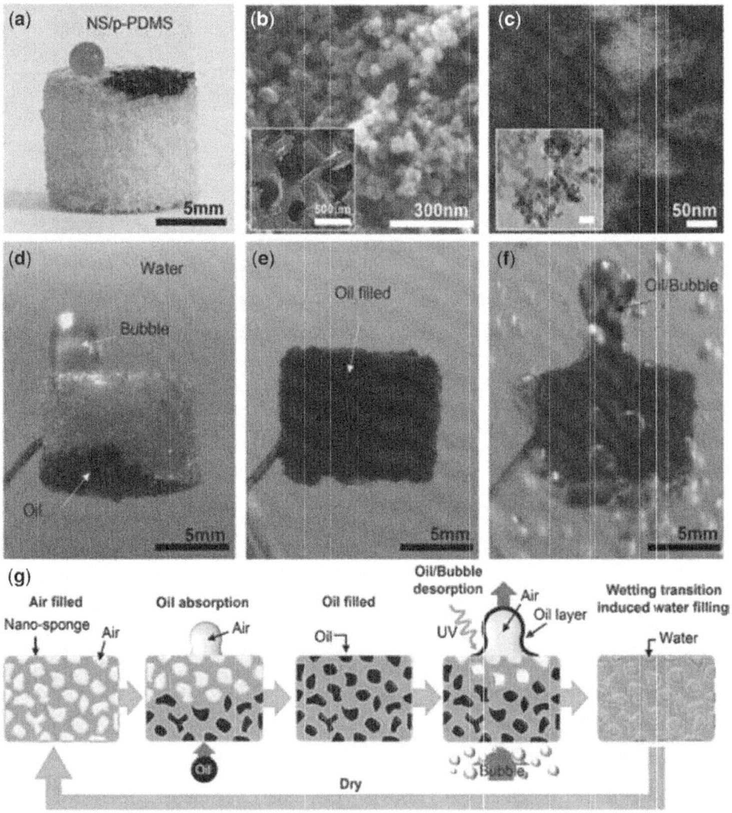

Figure 5.7 (a) Fabricated nano-sponge/porous PDMS (NS/p-PDMS) by dip coating. (b) SEM image of the surface structure of the NS/p-PDMS nano-sponge. (c) Compositional distribution: C (red), O (blue), Ti (green). The inset shows a bright field image of the nano-sponge. (Scale bar = 50 nm) (d) Underwater crude oil absorption of the nano-sponge, (e) fully absorbed state, and (f) oil desorption with air bubble flow/UV irradiation. (g) A schematic procedure of oil absorption and desorption with UV-responsive wettability and air bubble flow. Reproduced with permission from (Kim *et al.* Copyright 2015) Nature Publishing Group.

5.3.3 Gas-regulated oil-water separation

Using gas to trigger the transition of the surface wettability is convenient, and recently the controllable oil-water separation by using gas treatment has been reported. Xu *et al.* (2015) showed a superamphiphobic coating on fabric with an ammonia-triggered transition to superhydrophilic and superoleophobic for oil–water separation. The coating was prepared from a mixture of silica nanoparticles and heptadecafluorononanoic acid (HFA)-modified TiO_2 sol by a facile dip-coating method. Due to the low surface energy of the HFA and both the micro-scale roughness of the fabric weave structure and the nanoscale roughness of silica NPs, the coating exhibited superamphiphobicity (both superhydrophobicity and superoleophobicity) in air. When exposed to ammonia vapor, the coated fabric turned superhydrophilic within three seconds, but was still highly oil repellent (Figure 5.8b and 5.8c). The superhydrophobic to superhydrophilic transition of the ammonia treated fabric can be attributed to the formation of ammonium carboxylate ions in the presence of ammonia by cleavage of titanium carboxylate coordination bonding. With such an ammonia triggered transition of the wettability, the coated fabric can be used for the controllable oil-water separation process as demonstrated in Figure 5.8d. Due to the superamphiphobicity of the coated fabric, both hexadecane and water were retained at the beginning. However, upon the introduction of ammonia into the lower tube through a syringe needle, the fabric allowed water penetration but retains hexadecane, realizing the controlled oil-water separation.

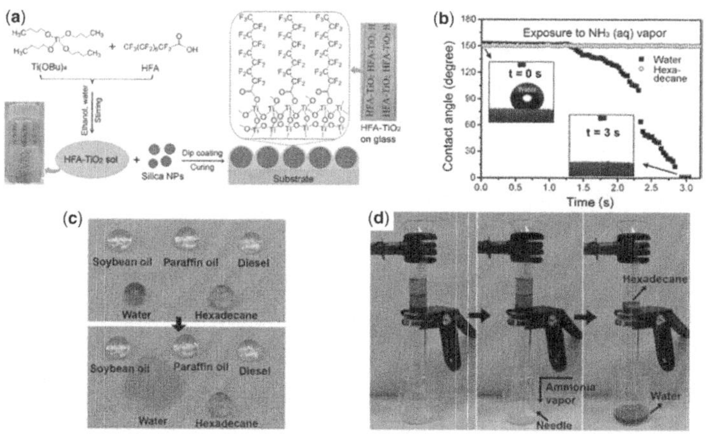

Figure 5.8 (a) Schematic illustration of the preparation procedure for superamphiphobic coating. (b) Time-dependence of contact angle for a water droplet and a hexadecane droplet placed on the coated polyester fabric upon exposure to ammonia vapor. (c) Photos of liquid droplets on the coated polyester fabric before (top panel) and after (bottom panel) ammonia exposure. (d) Controlled oil–water separation by the functionalized polyester fabric. Reproduced with permission from (Xu *et al.* Copyright 2015) John Wiley and Sons.

CO_2 gas recently has been also employed as a trigger to achieve the controllable oil-water separation. In a recent study of Che *et al.* (2015), smart nanostructured electrospun polymer membranes which were capable of switching oil/water wettability using CO_2 as the trigger were developed, and highly controlled oil/water separation using CO_2 as a trigger was reported (Figure 5.9). In this study, the copolymer of polymethylmethacrylate-co-poly(N,N-dimethylaminoethyl methacrylate) (PMMA-co-PDEAEMA) was first prepared by radical copolymerization reactions, and then was electrospun into continuous polymer nanofibers (Figure 5.9a and 5.9b). In the absence of CO_2, the nanofiber membrane exhibited highly hydrophobic and underwater superoleophilic properties. However, when the membrane was treated with CO_2 gas, a significant transition of the membrane from hydrophobicity to hydrophilicity was observed, and remarkably the oil wettability of the membranes was totally reversed to an underwater superoleophobic property after bubbling CO_2 into the aqueous solution (Figure 5.9c). It was also found that CO_2 could effectively adjust the level of the roughness and oil/water wettability of the nanofibers, and showed that the switching process allowed for the switchable oil-water separation (Figure 5.9d).

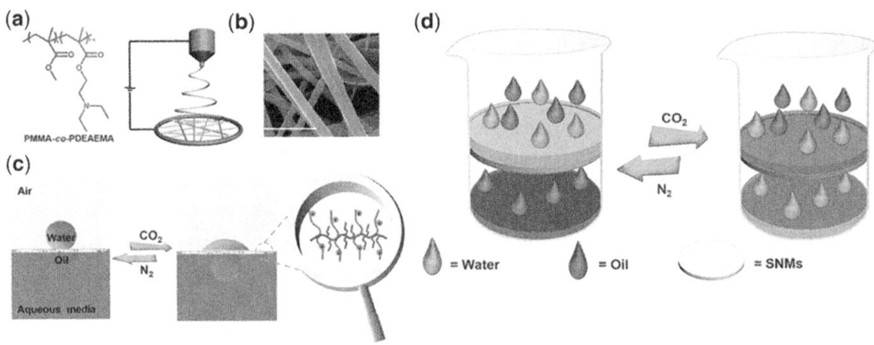

Figure 5.9 Fabrication process and CO_2 switchable oil/water wettability of the smart nanostructured membranes. (a) Diagram of the electrospinning process for fabricating PMMA-co-PDEAEMA nanofibers. (b) SEM images of a representative porous membranes with a diameter of 700 nm. Scale bars: 5 mm. (c) Illustration of the CO_2 switchable oil/water wettability between hydrophobicity/oleophilicity and hydrophilicity/oleophobicity. (d) Representation of the CO_2 switchable oil/water on–off switch. Reproduced from with permission from (Che *et al.* Copyright 2015) John Wiley and Sons.

5.3.4 Temperature controlled oil-water separation

Controllable oil-water separation has also been reported by using temperature as a trigger. Polyisopropylacrylamide (PNIPAM) is a comprehensively studied thermal-responsive polymer that exhibits an extended hydrophilic chain conformation below its lower critical solution temperature (LCST) in aqueous solution, and undergoes a phase transition to an insoluble and hydrophobic aggregate above its LCST. Sun *et al.* (2004) first reported the reversible switch between superhydrophobicity and superhydrophilicity of a surface in air by using the PNIPAM brushes. Then the underwater wettability and adhesion of oil to a thermal-responsive PNIPAM hydrogel was reported by Chen and co-workers (2010). The hydrogel surface could switch from a superoleophobic and low adhesive state below the LCST of PNIPAM to an oleophobic and high adhesive state above its LCST. By using a PNIPAM containing block copolymer, Xue *et al.* (2013) reported a temperature controllable dual water/oil on–off switch, and realized the temperature-controlled oil-water separation (Figure 5.10). In this study, a PMMA-b-PNIPAM block copolymer was selected, in which the PMMA is hydrophilic and oleophilic under water. PMMA has a high glass transition temperature, and can act as the physical crosslink of PNIPAM. Therefore, PMMA and PNIPAM nanodomains self-assembled by PMMA-b-PNIPAM exhibited unique wettability properties when the temperature changes, due to their synergistic effect (Figure 5.1c).

Cao *et al.* (2014) developed a thermo and pH doubly controllable oil/water separation material by coating thermo and pH dual-responsive poly (dimethylamino) ethyl methacrylate (PDMAEMA) hydrogel on stainless steel mesh substrates. They showed that when the temperature was greater than 55°C or pH was higher than 13, oil could permeate through the mesh since the water retention capacity of PDMAEMA hydrogel was significantly reduced and the swelling volume was decreased (Figure 5.11). The change in water retention capacity and volume was the macroscopic performance of transition of intra- and intermolecular hydrogen bonds in PDMAEMA hydrogel. Depending on the dual-responsive characteristic, oil and water can sequentially permeate through the mesh, and thus *in situ* collection of the oil and water can be achieved by adjusting temperature and pH.

By utilizing photothermal effect of gold (Au) nanorods and thermo-responsive PNIPAM, Hu *et al.* (2015) reported photothermal-responsive single-walled carbon nanotube-based ultrathin membranes for on/off switchable separation of oil-in-water nanoemulsions (Figure 5.12). In this study, photothermal-responsive ultrathin Au nanorods/poly(N-isopropylacrylamide-co-acrylamide) (PNIPAM-co-PAM) cohybrid single-walled carbon nanotube (SWCNT) nanoporous membranes were constructed. Such membranes are capable of separating oil-in-water nanoemulsions because they feature hydrophilicity, underwater oleophobicity, and nanometer pore sizes. It is remarkable that the permeation flux can be simply modulated by light illumination during the process of separation, due to the incorporation of thermal-responsive copolymers and Au nanorods.

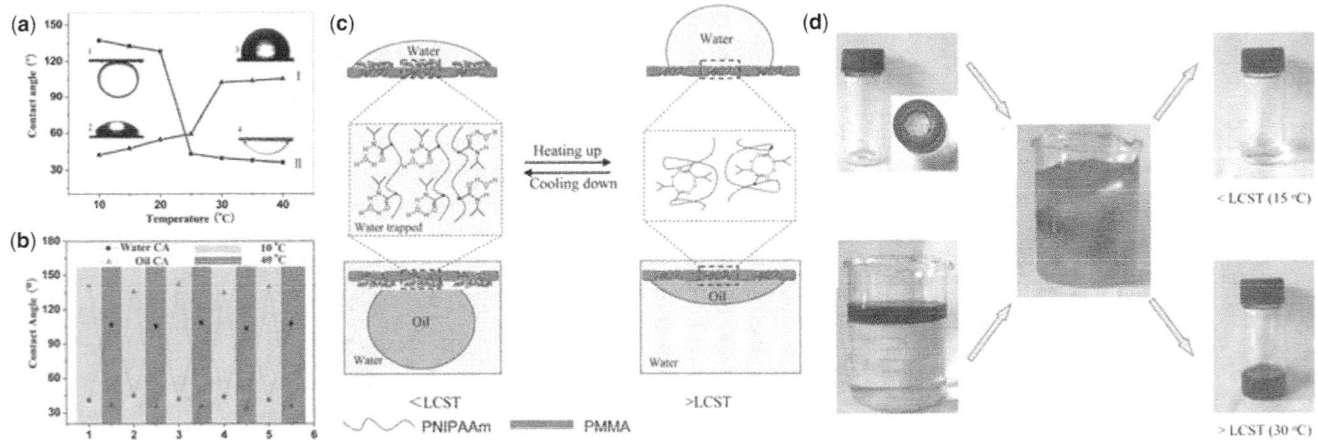

Figure 5.10 (a) Temperature dependences of water and oil CAs for a PMMA-b-PNIPAM film. (b) Reversible water and oil CA transition of the block copolymer film at different temperatures indicating excellent reproducibility and stability. (c) Diagram of reversible formation of intermolecular hydrogen bonding between PNIPAM chains and water molecules below the LCST, which leads to hydrophilicity/oleophobicity, and intramolecular hydrogen bonding between C=O and N—H groups in PNIPAM chains above the LCST, which leads to hydrophobicity/oleophilicity. Reproduced with permission from (Xue *et al.* Copyright 2013) John Wiley and Sons.

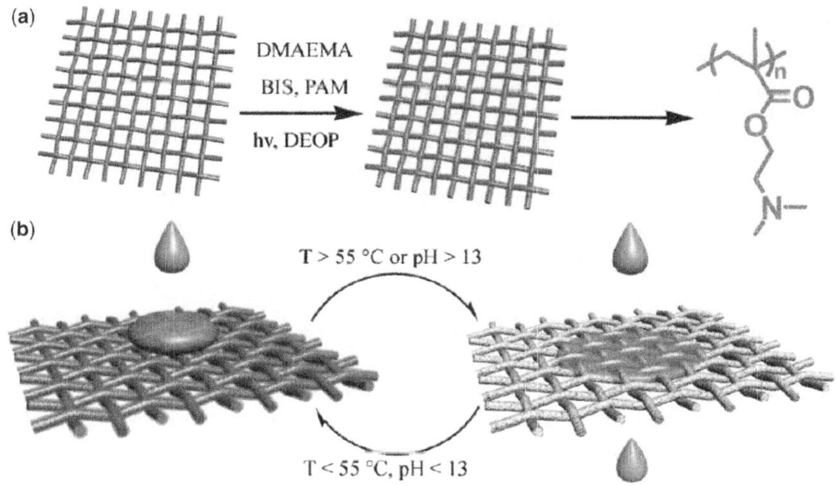

Figure 5.11 Schematic description of the preparation of PDMAEMA hydrogel coated mesh and the opposite wettability of the mesh when contacted with oil. (a) The stainless steel mesh as substrates was coated with PDMAEMA hydrogel which polymerized by DMAEMA monomer, N,N′-methylene bisacrlamide (BIS) as the chemical crosslinker, and polyacrylamide (PAM) as thickener under UV light. (b) Under 55°C (pH 7) and pH lower than 13, oil cannot permeate though the mesh, whereas when the temperature is above 55°C or pH is larger than 13, oil can pass through the mesh easily. Reproduced with permission from (Cao *et al.* Copyright 2014) American Chemical Society.

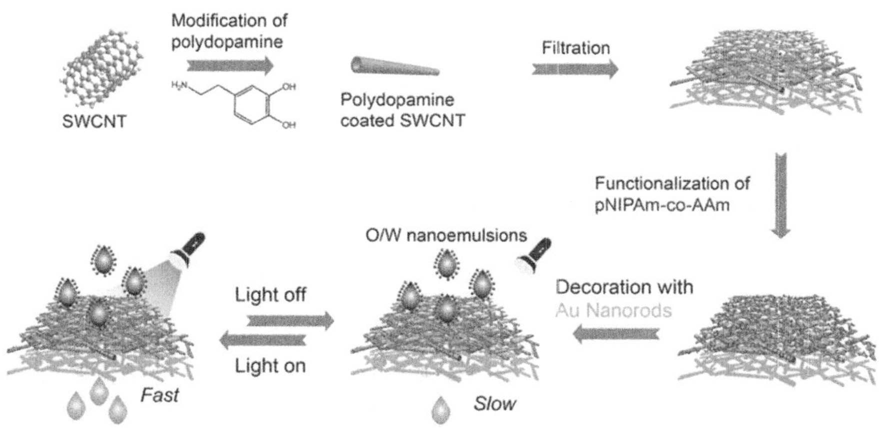

Figure 5.12 Schematic illustration for the fabrication of photothermal-responsive Au nanorods/PNIPAM-co-PAM cohybrid SWCNT ultrathin membranes. Reproduced with permission from (Hu *et al.* Copyright 2015) American Chemical Society.

5.3.5 Solvent-manipulated and ion-exchange controllable oil-water separation

Through a facile solvent exchange, surface property of the separation membrane could be changed, and a solvent-manipulated controllable oil-water separation could be achieved. Liu *et al.* (2014) prepared superhydrophilic and underwater superoleophobic Cu mesh with well-distributed copper hydroxide nanoneedle arrays by chemical oxidation (Figure 5.13a and 13b). The microsized pores in combination with nanoscale roughness endowed a hierarchical composite structure to the mesh t. When the mesh was immersed in an ethanol solution of stearic acid for 5 min, the surface can be converted to superhydrophobic and undertwater superoleophilic. After the stearic acid absorbed mesh was immersed in tetrahydrofuran (THF) for a short time, the superhydrophilicity could be recovered by the desorption of the stearic acid. More importantly, the reversible superhydrophobic-and-superhydrophilic switch can be repeated multiple times with almost no visible morphology variation.

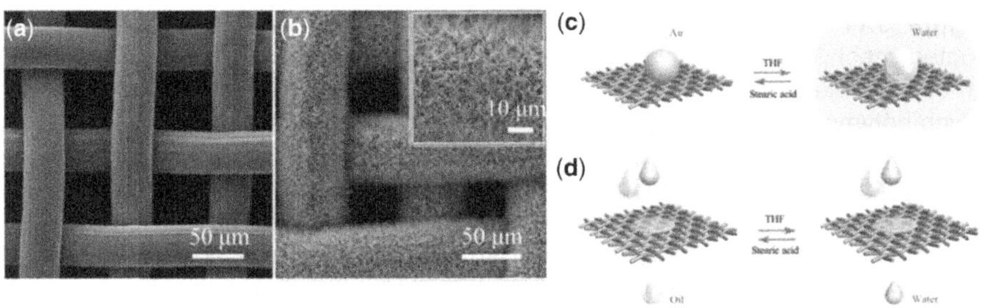

Figure 5.13 SEM images of (a) the untreated copper mesh with a smooth surface; (b) the oxidized mesh covered with $Cu(OH)_2$ nanoneedle arrays; the inset is the high-magnification view of nanoneedle arrays; (c, d) Schematic diagrams of the as-prepared mesh with reversible wettability switching for controllable oil/water separation: (c) When dipped into a stearic acid ethanol solution for 5 min, the mesh is superhydrophobic, whereas it converts to superhydrophilic and underwater to superoleophobic by immersion in THF for 5 min. (d) After it is modified with stearic acid, oil can penetrate through the mesh, but water cannot, whereas, when treated by THF, water can pass through the mesh quickly, but oil cannot. Reproduced with permission from (Liu *et al.* Copyright 2014) American Chemical Society.

On the other hand, the surface functional groups of superwetting membranes can also be tuned by using an ion-exchange method to realize the controllable oil-water separation. Zhang *et al.* (2014) reported an ion-exchange method to switch superhydrophobicity and superhydrophilicity of surface. In this work, they first prepared a rough surface on a stainless steel mesh substrate by a sol–gel process, and then a subsequent polyelectrolyte multilayer (PEM) coating was deposited on the surface. The surface energy of the PEM coating can be controlled by the

PEM counterions, and these counterions are easily ion exchanged, affording switchable separation and facile mesh rejuvenation. Most significantly, designed and constructed appropriately, these meshes can selectively separate either oil or water from their mixtures.

5.3.6 Electric field tuned oil-water separation

Electric field has been explore as a way of tuning solid surface wettability as well as for controllable oil-water separation process. Liu *et al.* (2011) first reported reversible underwater switching between superoleophobicity and superoleophilicity on conducting polymer nanotube arrays by using conducting polypyrrole (PPy) nanotube arrays, which were synthesized by a template method followed by electrochemical polymerization (Figure 5.14). A reversible superoleophobicity to superoleophilicity transition was controlled by tuning the electrochemical driving potential, which was attributed to the cooperation of electrochemical tunable doping, electrical double layer and alignment of nanostructures.

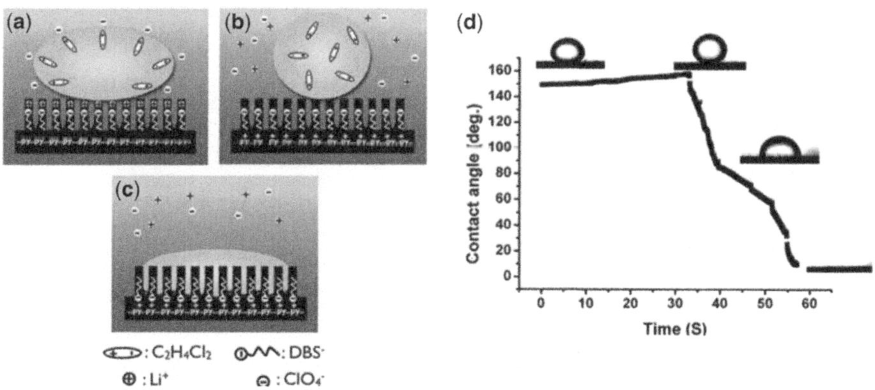

Figure 5.14 Mechanism of wetting states of oil drop on PPy nanotube arrays in different redox states. (a) In reduced state. (b) At the early stage of oxidation. (c) Final stage of oxidation. (d) Contact angle vs. time for an oil droplet on PPy nanotube arrays at oxidizing potential proving the real-time transition from superoleophobicity to superoleophilicity. Reproduced with permission from (Liu *et al.* Copyright 2011) Royal Society of Chemistry.

An electric field provides a facile route for tuning the wettability of polar (or conducting) liquids. The decrease in the macroscopic contact angle for a sessile polar liquid droplet on a dielectric in response to an electric field is commonly known as electrowetting on a dielectric. By utilizing electrowetting mechanism of polar liquid (water) and the non-responsive of non-polar liquids (oils) under electric field, an on-demand oil-water separation in response of eletric field was reported by Kwon *et al.* (Figure 5.15) (2012). Such on-demand separation can allow for the remote operation and automation of oil-water separation units. Besides oil-water separation, it has other potential applications such as designing microfluidic valves that selectively allow one liquid to flow through while retaining the other.

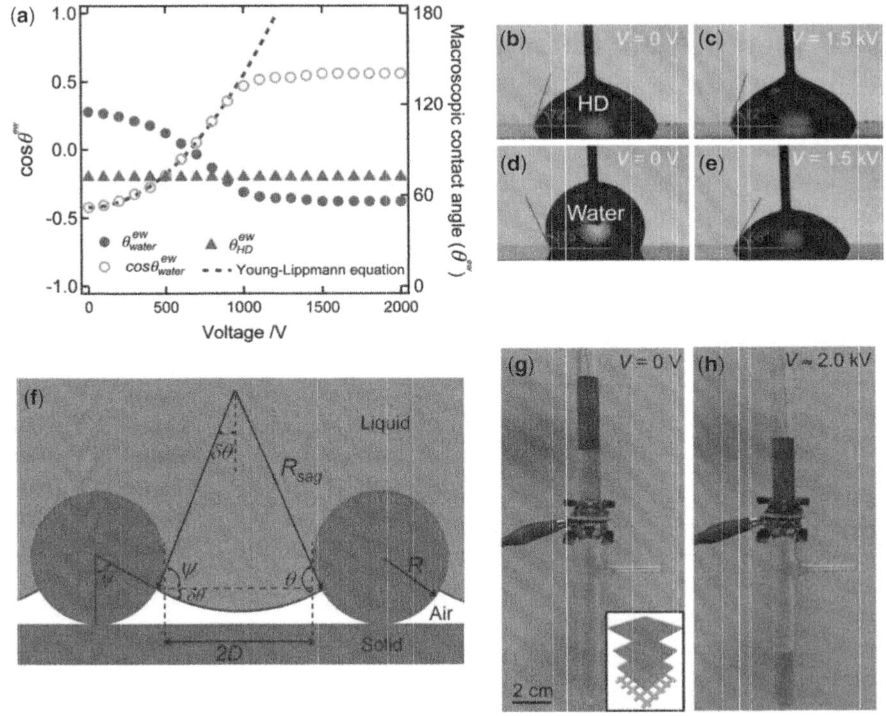

Figure 5.15 (a) Macroscopic contact angles for water and hexadecane (HD) as a function of applied voltage on the non-textured substrate. (b, c) The macroscopic contact angle for hexadecane remains unchanged with increasing voltage. (d, e) The macroscopic contact angle for water decreases with increasing voltage. (f) A schematic showing the pressure-induced sagging of the liquid-air interface. (g) An apparatus with a liquid column of free oil (dyed red) and water (dyed blue) above the membrane before applying an electric field. The inset shows a schematic of the membrane module. (h) Water permeates through while hexadecane is retained above the membrane when a voltage $V \approx 2.0$ kV is applied. Reproduced with permission from (Kwon *et al.* Copyright 2012) John Wiley and Sons.

5.4 SUMMARY AND PERSPECTIVE

In this chapter, we have presented the recent advances in the field of controllable oil-water separation based on responsive superwetting membranes, which provide great opportunity in effective oil spill response. However, although some achievements have been achieved in this area, there are still many challenges. First, most of the current responsive membranes in terms of the switchable superhydrophobicity-superoleophobicity and superhydrophilicity-underwater superoleophobicity are realized *ex situ*, i.e., additional treatments (drying or heating) are generally required to achieve a full cyclic switching of the wettability. A truly *in situ* switch of the superhydrophobicity-superoleophobicity and superhydrophilicity-underwater superoleophobicity in water or in oil system is still difficult to obtain. Secondly, it is still a big challenge to prepare these responsive materials at a large scale due to the high preparation cost. Thirdly, the mechanical stability as well as the multiple responsiveness of the responsive membrane materials need to be improved to satisfy practical applications. Having said that, we believe that with the creation of new knowledge in colloidal and interfacial science as well as the fast development of novel and muti responsive polymers especially block copolymers, many breakthroughs can be expected in the near future. Furthermore, we believe the responsive superwetting membranes also hold great potential in other oil-water related applications, such as microfluidic micro-nanosized droplet manipulation, antibiofouling, and self-cleaning coatings.

5.5 REFERENCES

Asatekin A. and Mayes A. M. (2009). Oil industry wastewater treatment with fouling resistant membranes containing amphiphilic comb copolymers. *Environmental Science and Technology*, **43**(12), 4487–4492.

Bai Z., Wang H. and Tu S.-T. (2011). Oil–water separation using hydrocyclones enhanced by air bubbles. *Chemical Engineering Research & Design*, **89**(1), 55–59.

Cao Y., Liu N., Fu C., Li K., Tao L., Feng L. and Wei Y. (2014). Thermo and pH dual-responsive materials for controllable oil/water separation. *ACS Applied Materials & Interfaces*, **6**(3), 2026–2030.

Cassie A. B. D. and Baxter S. (1944). Wettability of porous surfaces. *Transactions of the Faraday Society*, **40**, 546–551.

Chan Y. J., Chong M. F., Law C. L. and Hassell D. G. (2009). A review on anaerobic–aerobic treatment of industrial and municipal wastewater. *Chemical Engineering Journal*, **155**(1), 1–18.

Che H., Huo M., Peng L., Fang T., Liu N., Feng L., Wei Y. and Yuan J. (2015). CO_2-responsive nanofibrous membranes with switchable oil/water wettability. *Angewandte Chemie International Edition*, **54**(31), 8934–8938.

Chen L., Liu M. J., Lin L., Zhang T., Ma J., Song Y. L. and Jiang, L. (2010). Thermal-responsive hydrogel surface: tunable wettability and adhesion to oil at the water/solid interface. *Soft Matter*, **6**(12), 2708–2712.

Cheng Z., Wang J., Lai H., Du Y., Hou R., Li C., Zhang N. and Sun K. (2015). pH-controllable on-demand oil/water separation on the switchable superhydrophobic/superhydrophilic and underwater low-adhesive superoleophobic copper mesh film. *Langmuir*, **31**(4), 1393–1399.

Chu Z., Feng, Y. and Seeger S. (2015). Oil/water separation with selective superantiwetting/superwetting surface materials. *Angewandte Chemie International Edition*, **54**(8), 2328–2338.

Drioli E. and Giorno L. (2010). 1st edn, Comprehensive Membrane Science and Engineering. Elsevier Science.

Dubansky B., Whitehead A., Miller J. T., Rice C. D. and Galvez F. (2013). Multitissue molecular, genomic, and developmental effects of the deepwater horizon oil spill on resident Gulf Killifish (Fundulus grandis). *Environmental Science and Technology*, **47**(10), 5074–5082.

El-Kayar A., Hussein M., Zatout A. A., Hosny A. Y. and Amer A. A. (1993). Removal of oil from stable oil-water emulsion by induced air flotation technique. *Separations Technology*, **3**(1), 25–31.

Feng L., Zhang Z., Mai Z., Ma Y., Liu B., Jiang L. and Zhu D. (2004). A super-hydrophobic and super-oleophilic coating mesh film for the separation of oil and water. *Angewandte Chemie International Edition*, **43**(15), 2012–2014.

Gao S. J., Shi Z., Zhang W. B., Zhang F. and Jin J. (2014). Photoinduced superwetting single-walled carbon nanotube/TiO_2 ultrathin network films for ultrafast separation of oil-in-water emulsions. *ACS Nano*, **8**(6), 6344–6352.

He K., Duan H., Chen G. Y., Liu X., Yang W. and Wang D. (2015). Cleaning of oil fouling with water enabled by Zwitterionic polyelectrolyte coatings: overcoming the imperative challenge of oil–water separation membranes. *ACS Nano*, **9**(9), 9188–9198.

Head I. M., Jones D. M. and Röling W. F. M. (2006). Marine microorganisms make a meal of oil. *Nature reviews Microbiology*, **4**, 173–182.

Hu L., Gao S., Ding X., Wang D., Jiang J., Jin J. and Jiang L. (2015). Photothermal-responsive single-walled carbon nanotube-based ultrathin membranes for on/off switchable separation of oil-in-water nanoemulsions. *ACS Nano*, **9**(5), 4835–4842.

Ju G., Cheng M. and Shi F. (2014). A pH-responsive smart surface for the continuous separation of oil/water/oil ternary mixtures. *NPG Asia Materials*, **6**, e111.

Jung Y. C. and Bhushan B. (2009). wetting behavior of water and oil droplets in three-phase interfaces for hydrophobicity/philicity and oleophobicity/philicity. *Langmuir*, **25**(24), 14165–14173.

Kajitvichyanukul P., Hung Y.-T. and Wang L. (2006). Oil water separation. In Advanced Physicochemical Treatment Processes, L. K. Wang, Y.-T. Hung and N. K. Shammas (eds.), Humana Press, pp. 521–548.

Kim D. H., Jung M. C., Cho S.-H., Kim S. H., Kim H.-Y., Lee H. J., Oh K. H. and Moon M.-W. (2015). UV-responsive nano-sponge for oil absorption and desorption. *Scientific Reports*, **5**, 12908.

Kwon G., Kota A., Li Y., Sohani A., Mabry, J. M. and Tuteja, A. (2012). On-demand separation of oil-water mixtures. *Advanced Materials*, **24**(27), 3666–3671.

Li J.-J., Zhou Y.-N., and Luo Z.-H. (2015). Smart fiber membrane for pH-induced oil/water separation. *ACS Applied Materials & Interfaces*, **7**(35), 19643–19650.

Liu M., Wang S., Wei Z., Song Y. and Jiang L. (2009). Bioinspired design of a superoleophobic and low adhesive water/solid interface. *Advanced Materials*, **21**(6), 665–669.

Liu M., Liu X., Ding C., Wei Z., Zhu Y. and Jiang L. (2011). Reversible underwater switching between superoleophobicity and superoleophilicity on conducting polymer nanotube arrays. *Soft Matter*, **7**(9), 4163–4165.

Liu N., Cao Y., Lin X., Chen Y., Feng L. and Wei Y. (2014). A facile solvent-manipulated mesh for reversible oil/water separation. *ACS Applied Materials & Interfaces*, **6**(15), 12821–12826.

McNutt M. K., Camilli R., Crone T. J., Guthrie G. D., Hsieh P. A., Ryerson T. B., Savas O. and Shaffer F. (2012). Review of flow rate estimates of the Deepwater Horizon oil spill.

Proceedings of the National Academy of Sciences of the United States of America, **109**(50), 20260–20267.

Mostefa N. M. and Tir M. (2004). Coupling flocculation with electroflotation for waste oil/water emulsion treatment. Optimization of the operating conditions. *Desalination*, **161**(2), 115–121.

Nordvik A. B., Simmons J. L., Bitting K. R., Lewisa A. and Kristiansen T. S. (1996). Oil and water separation in marine oil spill clean-up operations. *Spill Science & Technology Bulletin*, **3**(3), 107–122.

Sahoo B. N. and Kandasubramanian B. (2014). Recent progress in fabrication and characterization of hierarchical biomimetic superhydrophobic structures. *RSC Adv.*, **4**(42), 22053–22093.

Short J. W., Rice S. D., Heintz R. A., Carls M. G. and Moles A. (2003). Long-term effects of crude oil on developing fish: lessons from the Exxon Valdez oil spill. *Energy Sources*, **25**(6), 509–517.

Sun T., Wang G., Feng L., Liu B., Ma Y., Jiang L. and Zhu D. (2004). Reversible switching between superhydrophilicity and superhydrophobicity. *Angewandte Chemie International Edition*, **43**(3), 357–360.

Sun T. L., Feng L., Gao X. F. and Jiang L. (2005). Bioinspired surfaces with special wettability. *Accounts of Chemical Research*, **38**(8), 644–652.

Tian D., Zhang X., Tian Y., Wu Y., Wang X., Zhai J. and Jiang L. (2012). Photo-induced water–oil separation based on switchable superhydrophobicity–superhydrophilicity and underwater superoleophobicity of the aligned ZnO nanorod array-coated mesh films. *Journal of Materials Chemistry* **22**(37), 19652–19657.

Wang B., Liang W., Guo Z. and Liu W. (2015). Biomimetic super-lyophobic and super-lyophilic materials applied for oil/water separation: a new strategy beyond nature. *Chemical Society Reviews* **44**(1), 336–361.

Wenzel R. N. (1936). Resistance of solid surfaces to wetting by water. *Industrial and Engineering Chemistry*, **28**(8), 988–994.

Xu Z., Zhao Y., Wang H., Wang X. and Lin T. (2015). A superamphiphobic coating with an ammonia-triggered transition to superhydrophilic and superoleophobic for oil–water separation. *Angewandte Chemie International Edition*, **54**(15), 4527–4530.

Xue B., Gao L., Hou Y., Liu Z. and Jiang L. (2013). Temperature controlled water/oil wettability of a surface fabricated by a block copolymer: application as a dual water/oil on–off switch. *Advanced Materials*, **25**(2), 273–277.

Xue Z., Wang S., Lin L., Chen L., Liu M., Feng L. and Jiang L. (2011). A novel superhydrophilic and underwater superoleophobic hydrogel-coated mesh for oil/water separation. *Advanced Materials*, **23**(37), 4270–4273.

Xue Z., Cao Y., Liu N., Feng L. and Jiang L. (2014). Special wettable materials for oil/water separation. *Journal of Materials Chemistry A*, **2**(8), 2445–2460.

Young T. (1805). An essay on the cohesion of fluids. *Philosophical Transactions of the Royal Society of London*, **95**, 65–87.

Zhang G., Li M., Zhang B., Huang Y. and Su Z. (2014). A switchable mesh for on-demand oil–water separation. *Journal of Materials Chemistry A*, **2**(37), 15284–15287.

Zhang L., Zhang Z. and Wang, P. (2012). Smart surfaces with switchable superoleophilicity and superoleophobicity in aqueous media: toward controllable oil/water separation. *NPG Asia Materials*, **4**, e8.

Zhang L., Zhong, Y., Cha, D. and Wang P. (2013). A self-cleaning underwater superoleophobic mesh for oil-water separation. *Scientific Reports*, **3**, 2326.

Zhu Y., Zhang F., Wang D., Pei X. F., Zhang W. and Jin, J. (2013). A novel zwitterionic polyelectrolyte grafted PVDF membrane for thoroughly separating oil from water with ultrahigh efficiency. *Journal of Materials Chemistry A A*, **1**(18), 5758–5765.

Zhu Y., Wang D., Jiang L. and Jin J. (2014). Recent progress in developing advanced membranes for emulsified oil/water separation. *NPG Asia Materials*, **6**, e101.

Chapter 6

Design of the next-generation FO draw solution

Aaron D. Wilson

6.1 INTRODUCTION

6.1.1 History of forward osmosis draw solutes

Engineered osmosis (EO) or osmotically driven membrane processes (ODMP) involves the transfer of water from a feed solution across a semi-permeable membrane to a draw solution with a higher osmotic concentration, Figure 6.1.

If the target product of an EO process is water, diluted draw, or concentrated feed the process is generally referred to as forward osmosis (FO) (Cath *et al.* 2006; Chung *et al.* 2012; Hoover *et al.* 2011). If energy is extracted from the water transfer then the process is pressure retarded osmosis (PRO) or in a more specific incarnation an osmotic heat engine (OHE) (Achilli & Childress, 2010). FO's roots can be traced back to water-selective "dialysis", a process used to remove water from organic solvent published as early as 1932 (Schaack, 1932). PRO as a method to generate useful energy was identified in 1954 (Pattle, 1954). The Ammonia-CO_2 draw system which would become so significant in modern ODMP research was first identified in 1964 by Neff (Neff, 1964). Despite these and other efforts in ODMP, the publications in the field were intermittent and sporadic until a series of papers starting in 2005 reintroduced a better developed variation on the ammonia–CO_2 draw system (McCutcheon *et al.* 2005). Since 2005 there has been an exponential growth in publications addressing ODMP in terms of draw solute development, membrane development, and theoretical considerations.

A draw solute is a material dissolved in water to supply the osmotic concentration necessary to transfer water out of a feed solution with a lower osmotic concentration. The role of the draw solute in sparking the current interest in ODMP highlights their importance, Figure 6.2. A number of reviews already exist which are partially or completely dedicated to draw solutes (Achilli *et al.* 2010; Chekli *et al.* 2012; Ge *et al.* 2013; Li & Wang, 2013; Shaffer *et al.* 2015; Qasim *et al.* 2015). Common to these reviews is a list of desirable characteristic for draw solutes which usually includes a high osmotic concentration (thermodynamic driving force), favorable mass transport properties (low viscosity, high diffusivity, solution mixing, membrane specific interactions, water flux), membrane impermeability, effective means of recovery, nontoxicity, and low cost.

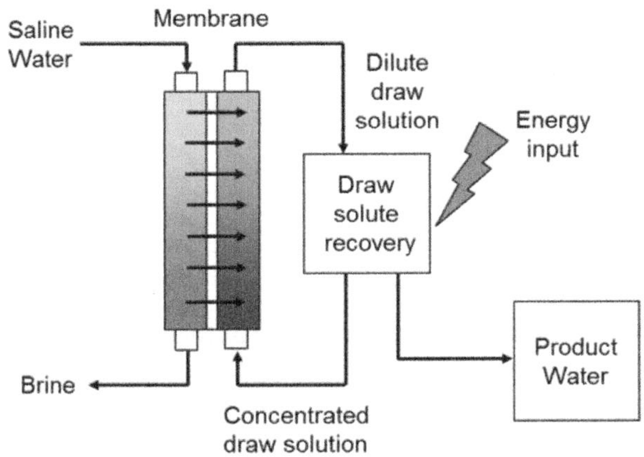

Figure 6.1 Simplified depiction of the FO process.

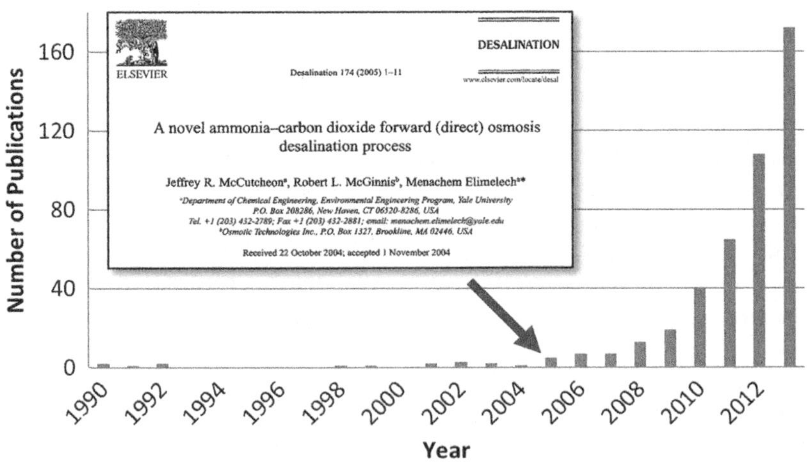

Figure 6.2 All references resulting from a topic search for "forward osmosis" or "pressure retarded osmosis" in Web of Science. Inset (McCutcheon *et al.* 2005).

In terms of application FO has been explored as a pretreatment process for reverse osmosis (RO) and as a primary water treatment technology when using stimuli responsive draw solutes. With RO approaching thermodynamic limits in terms of treatable concentrations and energy efficiency FO processes have been considered as a promising alternative to reduce cost. Despite suggestion and innuendo, FO is not expected improve on the energy consumption of RO. The thermodynamic cost of desalinating a solution is determined by its solutes. Modern is RO highly optimized and is highly effective treating solution between brackish and sea water osmotic concentrations. As the operational limitation and benefits of FO have been better identified, the focus has shifted away from developing general desalination to developing niche applications which best capitalize on FO's advantages. FO has the potential to reduce water treatment lifecycle costs and address specific water treatment needs in three ways. First, FO may reduce pretreatment costs; pretreatment costs can be greater than the "core" process energy costs in many applications. Second, FO can treat higher concentrations than RO and obtain higher volume recoveries; this means handling (and treating) less feed water and producing a smaller volume of concentrated brine than RO. Third, FO with temperature responsive draw solutes or use of thermal recovery techniques such as membranes distillation may shift energy use from electricity to more cost effective thermal energy (or low-grade thermal energy). Unless FO serves exclusively as a pretreatment for RO or another osmotic filtration all draw solutes must 1) have a readily generated high osmotic concentration state to treat high concentrations feeds and/or 2) the draw solutes should be regenerated thermally.

Draw solutes are often divided into two categories. First are conventional solutes which are concentrated through conventional means of dewatering a solution containing dissolved solids such as RO or some form of distillation (Achilli *et al.* 2010). The second group is stimuli responsive draw solutes sometimes referred to as unconventional draw solutes or next generation draw solutes. Unconventional draw solutes separate from water by any means or combination of means that can be devised. As mentioned above if the process is to be more efficient than the separation must be achieved by energy more affordable than hydraulic pressure.

6.1.2 Recent trends in draw solutes

Every combination of "formally" monoatomic cations, ammoniums, halides, oxoanions, small organic cation/anions with less than 6 carbons, and common small neutral polar solutes (alcohols, sugars, ect.) have been considered for their use in ODMP by multiple research groups and industrial entities. There is value in many of these "simple" solutes but much of the research in draw solutes has moved to high molecular mass draw solutes. Advantages of high molecular mass draw solutes include dynamic methods of draw solute recovery and reduced membrane permeability. The challenges of high molecular mass draw solutes included reduced osmotic density, reduced maximum osmotic concentration, reduced carrying capacity, reduced osmotic density, mass transport issues, and fundamental material cost per osmotic unit (osmotic cost).

6.2 DESIGN OF DRAW SOLUTES

This text is dedicated to design of draw solutes rather than pure review draw solutes and focuses on the broad strategy of developing a draw solute. The method used evaluate system is three part, first simple technoeconomic thermodynamics analysis, second idealized kinetics analysis, and third practical concerns.

Does the proposed ODMP process have the thermodynamic requirements to improve on a state-of-the-art technology? If the thermodynamics of a process do not improve on the state-of-the-art based on liberal simplified lifecycle estimations than a draw solute and its associated system should be not be expected to ever be implemented profitably. This thermodynamic consideration is usually what divides academic curiosities from systems with the potential to lead to commercial products. In FO research the thermodynamics are often assumed, implied, or ignored. In some cases ignoring thermodynamics is a major error, no amount of optimization can bypass thermodynamic limits. Still skipping thermodynamic consideration can be an understandable error given the FO research communities focus on kinetics.

Kinetic, specifically membrane kinetics, occupies the majority of the focus in published literature around ODMP which could be expected given that the technology developed out of the membrane research community. Water flux across an FO membrane (J_w) is a rich area to study given that it depends on the membrane (thickness, porosity, tortuosity, hydrophilicity), draw solute (mass transport), solution concentrations (osmotic pressure), and operational conditions (temperature and pressure). Solute flux across the membrane (J_s) and rejection of solute in the feed solution provide additional membrane kinetics to study. Moving away from membrane kinetics there are system specific considerations such as draw solute separation and regeneration kinetics which vary depending on the draw solute. Most of the key kinetic values can be measured on lab-scale test equipment and extrapolated to full scale systems to predict the potential benefits of an FO process. While there is no defined minimum for the water flux in ODMP there is an often unstated understanding that many of the systems reported in the literature could never be cost effectively commercialized. In the cases where both the thermodynamic and kinetic metrics are met it may still be "impractical" to implement an ODMP system for other reasons.

Practical considerations include items such as current draw solute cost, draw solute degradation rate, material handling concerns, solute toxicity, and infrastructure costs. This is considered last because these issues can often be corrected, mitigated, or overcome during applied research efforts and in some cases do not emerge until process scale-up.

6.2.1 Physical properties of draw solute

6.2.1.1 Misconception with osmotic pressure

Osmotic pressure provides the thermodynamic driving force for ODMP processes (Wilson & Stewart, 2013). The first Nobel prize in chemistry was awarded to van't Hoff in 1901 for his work on osmotic pressure (van't Hoff, 1995) but despite this long history there is still confusion on the subject of osmotic pressure (Wang *et al.* 2014; Zhao, 2014). Three issues complicate working with osmotic pressure 1) popularity of non-proportional concentration units, 2) the use of thermodynamic software to calculated osmotic pressure, and 3) intrinsic problems with established electrolyte models.

Weight percent, total dissolved solids (tds) reported in parts per million (ppm), and molarity (*M*) are all non-proportional with osmotic pressure except at very low concentrations. Given that tds is the preferred units in the water treatment industry and molarity is the unit most often used by chemists it is easy to see how osmotic pressure became a problematic physical property to evaluate. Concentrations units which are more proportional to osmotic pressure include molality (mol/Kg), mole fractions, and solute/solvent mass fractions (Wilson & Stewart, 2013).

In the ODMP community the problems with electrolyte models have often been sidestepped by using thermodynamic software to calculate a solution's osmotic pressure. Calculations with this software have resulted in some dubious graphs featuring data points for materials beyond their solubility. In the case sucrose (343.3 g/mol) it is featured at 5 molar a concentration, well beyond its solubility of 2000 g/L or ~2 molar. This sort of mistake highlights the easy with which it is to miss use software to "calculate" flawed or impossible values. These mistakes happen even the when the public literature contains known empirically parameters. When software is used to calculate values for systems for which no empirical data is available it must be done with great care. This is especially important for process models which must address conditions over a range of concentrations. The energy required to remove a volatile solute from solution varies with concentration even if it is common to report values at specific values.

To describe the relationship between osmotic pressure and concentration requires an electrolyte model. Most electrolyte models are based on Debye-Hückel (DH) theory which emerged in the 1920's as an extension of work in particle physics. It is widely agreed that DH theory is "insufficient" to describe anything but a narrow range of conditions still the core theory has been assumed to be the best starting point to develop nearly all chemical physics description of electrolytes. Electrolyte models such as Davies equation, Pitzer equation, specific ion interaction theory (SIT theory), among many others start by modifying DH theory with additional terms to better match experimental results. Most these models include at least one concentration and temperature dependent coefficient to describe a solutes "activity". It is dubious that a theory intended to describe how a property changes with concentration contains an empirically derived parameter that also varies with concentration. In most cases these activity coefficients and related values are ultimately empirically derived fitting parameters. These issues with fundamental theory describing electrolyte concentration are important to remember as much of the work in FO moves to higher-concertation less-ideal feed and draw solutios.

To develop draw solutions for which the "activity" is unknown requires the adoption of an electrolyte model which correlates a solution's concentration with osmotic pressure. Our group has found the Morse equation to be both simple to use and useful across a wide range of concentrations.

$$\Pi = i\rho\left(\frac{n_s}{M_w - hn_sMW_w}\right)RT \tag{1}$$

This Morse expression (Equation 1) consists of the commonly recognized ideal gas constant (*R*) and temperature (*T*). Less universal is a concentration expressed here a solute hydration number (*h*) adjusted molality (moles solute/mass solvent), density correction (ρ), and a van't Hoff index, *i*.

Previously an experimentally determined density correction, ρ, was used to convert Kg to L (Wilson & Stewart, 2013) but it is unclear how that correction should be applied (McNally *et al.* 2015). The density correction is required to reconcile the molal concentration (mole/mass solute to mass solvent ratio) to the units used in the Morse equation. These units are based in the semi-empirical semi-theoretical nature of its predecessor the van't Hoff equation. The correction should convert the mass solute to either the volume of solvent or volume solution. If the correction is to the solvent volume than it would require the density of water or more likely the partial molar density of the solvent water is very close to water but impossible to separate from the partial molar density of the solute. If the conversion was to go from solvent to solution the corrected concentration would increase, in turn reducing van't Hoff index significantly. It is not clear which correction is correct and using the solvent density, the solvents partial molar density, or only the solution density but for most aqueous work a density correction of 1 Kg/L is a reasonable estimate given an empirical van't Hoff index.

The van't Hoff index is an empirically derived composite of information related to ion-pairing, waters of hydration, and a minor amount of information related to a density correction. Each of these components can be quantitatively addressed; the waters of hydration (which increases the uncorrected composite van't Hoff index), ion-pairing (which reduces the composite van't Hoff index), and density correction (which increases the uncorrected composite van't Hoff index). The van't Hoff index converts a molality to an osmolality. A general rule of thumb is that an Osm/kg can be converted into osmotic pressure with 24.8 (atm L/Osm).

Molality is not the ideal for polymers, dendrimers, nanoparticles, and other systems with large, variable, and/or averaged molecular masses. When the chemical composition of a material is known in term of monomers or number of functional groups it is often useful to consider normalized mole ratios ($i \times n/$ $Kg_{solvent}$) in which n is the mole quantity of a subunit such as carboxylate salt unit. Then the treatment is similar to that in Equation 1. For system not easily expressed by molality or subunit concentration it is reasonable to use a mass ratio of the solute to (M_s) to solvent (M_w) with an appropriately empirically derived index, i_m, to convert the units to Osm/kg, Equation 2. One of the beauties of Equation 2 is that works for both pure materials and mixtures and requires only the solute mass fraction (derivable from tds). This allows for complex evaluation process energies with virtually no information about chemical composition of the draw or feed solution (Reimund *et al.* 2015).

$$\Pi = i_m \left(\frac{M_s}{M_w} \right) RT \tag{2}$$

Various systems which will deviate from the expression found in Equations 1 and 2; for example, a decline from linearity due to solute association or limited "free" water at high concentrations. For such non-linear systems fitting with a viral expansion (polynomial) is a reasonable approach. This is theoretically more appropriate than fitting molar concentration data with a viral expansion without ever making the effort to plot the data against molality or mole ratios.

6.2.1.2 Maximum available osmotic pressure

The maximum osmotic pressure that can be practically obtained from a draw solute thermodynamically defines the feed solutions concentrations that can be treated. This defines the concentrated feed output and the fraction of water which can be removed. As stated above FO's ability to treat high concentration solution and obtain large water fraction a cardinal advantage that FO can have over RO. If the material is a common solute the "activity" at or near this concentration will be known but for a designer draw solute this "activity" will be an unknown. Directly measuring osmotic pressure is challenging and generally not worth the effort for developing ODMP. The most common methods to indirectly study osmotic pressure are freezing point depression and vapor pressure osmometry. Commercially available osmometers are generally limited to measuring concentrations below 4 Osm/Kg (<100 bar). It is highly desirable to have a draw solute with a maximum osmotic pressure higher than 100 bar. This value, π_{HI}, should be known or estimated for any proposed draw solute to understand process limitations.

For a counter current system the fraction of water obtained from the initial volume can be obtained from relationships featured in Equation 3. By normalized the feed solution input volume, V_{LI}, to unity it is possible to obtain the fraction of water removed for a given draw osmotic pressure, π_{HI}, for a given pressure differential or operating pressure, P_{op}, Equation 4. This fraction can be normalized to the available in the initial solution by dividing by the initial water fraction of the feed solution, w_{LI}, Equation 5.

$$\pi_{LO} = \left[\pi_{HI} - P_{op}\right] = \frac{\pi_{LI}}{\dfrac{V_{LI} + \Delta V / w_{LI}}{V_{LI}}} \tag{3}$$

$$\frac{\Delta V}{V_{LI}} = -\left(\left(\frac{\pi_{LI}}{\pi_{HI} - P_{op}}\right) - 1\right) w_{LI}^{-1} \tag{4}$$

$$\frac{\Delta V}{V_{LI} w_{LI}} = \frac{\Delta V}{V_{LIw}} \tag{5}$$

Extracting water from solutions that start at high concentrations or solutions that reach high concentrations during the removal of water require significant operating pressure, P_{op}, to maintain a reasonable water flux. It is not unreasonable to expect to need a 100 bar (~4 Osm/Kg) operating pressure to maintain flux. Plotting these conditions against hypothetical brackish, saline, and brine solutions, Figure 6.3, illustrates just how high an osmotic concentration a draw solution must provide to perform better than RO. In this scenario it takes a draw of ~6 Osm/Kg for a water fraction removal comparable to a seawater RO process. When scaled up (and possibly run in co-current FO cells) even more driving force may be required to maintain a useful flux and extraction fraction.

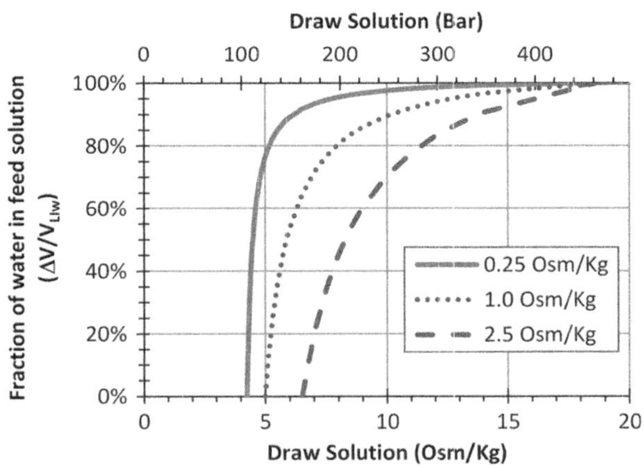

Figure 6.3 The fraction of available water in a feed solution which can be removed by a draw solution of a given osmotic pressure (assuming a 100 bar of excess osmotic pressure). Solid line "brackish water" 0.25 Osm/Kg (assuming 9,000 ppm TDS), dotted line "saline water" 1.0 Osm/Kg (assuming 35,000 ppm TDS), and dashed line "brine" 2.5 Osm/Kg (assuming 90,000 ppm TDS).

6.2.1.3 Entropic Sensitivity

Once diluted in a FO module the product water must be isolated and the draw solute generally needs to be recycled. As discussed above for a cost effective FO system (that is not a pretreatment process) it is expect that this separation will be thermally driven. Thermal energy will be used to drive the material between two equilibrium states. These equilibrium states can be imagined in terms of equilibrium constants K_{eq} or corresponding free energy value, ΔG. Whether ΔG has a positive or negative correlation with the aqueous solution osmotic concentration depends on the frame of reference. Regardless a thermal process the free energy sates can be further deconvoluted into enthalpy (ΔH) and entropy (ΔS) terms. The temperature sensitive portion of an equilibrium state is the entropic term, $T\Delta S$, Equation 6.

$$\Delta G = \Delta H - T\Delta S \tag{6}$$

It does not matter if the osmotic concentration increases or decreases with temperature what matter the magnitude of this change. This magnitude of change is a draw solutes osmotic entropic sensitivity and can be expressed by the quotient of the change osmotic concentration and change in temperature $\Delta C/\Delta T$ (Δ(Osm/Kg)/ΔK) or $\Delta\Pi/\Delta T$ (ΔBar/ΔK). Because in most cases the entire solution needs to be heated small temperature swings near the ambient temperature of the feed solution are desirable. A high entropic sensitivity suggests a stimuli responsive draw solute will be less costly to cycle requiring smaller swings in temperature to produce product water or regenerate the draw solution.

6.2.1.4 Minimum Stimuli-Driven Osmotic Concentration

The minimum stimuli-driven osmotic concentration (MSDOC) is the concentration that can be reached through piratical application of a stimulus to a stimuli responsive draw solute system. Stated plainly some amount of draw solute will often remain in the product water after a stimuli treatment. The stimuli can be the additional of heat, removal of heat, manipulation of a magnetic field, or some other process. The MSDOC solution may be ready for use depending on the application or may require further processing in other cases. Additional treatments could include osmotic filtration, membrane distillation, collative, biological, oxidative, stationary sorbent, ion-exchange, electrochemical, or some other process. There are many scenarios where an osmotic filtration has been considered. The trace stimuli responsive draw solute could concentrate of phase separate during the osmotic filtration. Either way the osmotic filtration will generally have the same thermodynamic energy requirements of an RO process even if it is NF or UF.

6.2.1.5 Carrying Capacity

There is only so much water than a draw solution can isolate with each pass through an FO module array before it is saturated. This carrying capacity is a major factor in terming the economic feasibility of a FO process. A higher carrying capacity means a smaller amount of draw solute needs to be cycle for each unit of product water. While discussing minimum stimuli-driven osmotic concentration it was mentioned that in general a stimuli must be applied to the entire dilute solution. Thus a higher carrying capacity also means a larger the portion of the solution is product water providing a smaller opportunity of parasitic heat energy losses.

The literature does not currently contain general methods to describe ideal or evaluate practical carrying capacity. The methods used here are related to our other work regarding the thermodynamics of ODMP (Reimund *et al.* 2015; Wilson & Stewart, 2013). An ideal counter-current FO apparatus a draw solution can be disrobed be described by Equation 7 in terms of osmotic pressure and Equation 8 in terms of concentration when those value are proportional to osmotic pressure. The carrying capacity of a solute can be defined as the water transferred divided by the volume of the draw solution (Equation 9) moles in the initial solution (Equation 10). To adapt Equation 9 and 10 to non-ideal conditions all that needs to be included is the empirically determined or selected values for C_{HI} and C_{HO}.

$$\pi_{HO} = \left[\pi_{LI} + P_{op}\right] = \frac{\pi_{HI}}{\frac{V_{HI} + \Delta V/w_{HI}}{V_{HI}}} \tag{7}$$

$$C_{HO} = \frac{C_{HI}}{\frac{V_{HI} + \Delta V/w_{HI}}{V_{HI}}} \tag{8}$$

$$\frac{\Delta V}{V_{HI}} = \left(C_{HI}/C_{HO} - 1\right) w_{HI} \tag{9}$$

$$K_{\Delta V} = \frac{\Delta V}{C_{HI} V_{HI}} = \frac{\left(C_{HI}/C_{HO} - 1\right)}{C_{HI}} \tag{10}$$

Hypothetically if the initial concentration of the draw solution (C_{HI}) is allowed to approach infinite the initial draw solution volume (V_{HI}) becomes negligible. Assuming infinite C_{HI} a thermodynamic limit can be defined by the concentration of a fully saturated draw solution $(C_{HO})^{-1}$, Equation 11. Because the C_{HO} can be described both practically and ideally it provides two theoretical limits which can be used to evaluate a draw solution performance when using a real initial draw solution concentration, C_{HI}.

$$\left(C_{HO}\right)^{-1} = \frac{\Delta V}{C_{HI} V_{HI}} \tag{11}$$

For low molecular mass draw solutes with high solubilities the carrying capacity of associated draw solutions are so similar for equivalent osmotic concentrations that in terms of application they are equivalent. The theoretical limits described by Equation 11 while not addressed in the literature was known in a practical sense for the groups that have been involved in scaling up FO process to industrial application. With the move to high molecular mass draw solutes there is a growing need to explicately describe carrying capacity and the solutes osmotic density as described in the next section.

6.2.1.6 Osmotic Density

Osmotic density is a simple concept that describes the osmotic pressure that can be generated per mass of draw solute. A solute's van't Hoff index can be divided by its molecular mass to generate its osmotic density that can be expressed in term of Osm/Kg or Bar/Kg, Equation 12, assuming a proportionality between osmotic pressure and concentration. Osmotic density allows the direct comparison of different draw solutes independent of a specific concentration. This osmotic density is useful for understanding the potential performance of draw with a high molecular mass. If the solution must reach 75 wt% before reaching 10 Osm/Kg (~250 bar) the draw solute will likely have significant mass transport issues and a limited carrying capacity.

$$\rho_\pi = \frac{i}{MW} = \frac{C_{\text{solute}}}{M_{\text{solute}}} \propto \frac{\pi_{\text{solute}}}{M_{\text{solute}}} \tag{12}$$

The osmotic density is a serious challenge for many high molecular mass draw solutes. For example, NaCl i (2.09 Osm/mol)/MW (0.05844 Kg/mol) has an osmotic density of $\rho_\pi = 35.8$ Osm/$\text{Kg}_{\text{solute}}$ it would require ~0.280 Kg of NaCl per Kg water to generate a 10 Osm/Kg solution (~22 wt%). For comparison 1-cyclohexylpiperidinium bicarbonate (CHP) has i (1.75 Osm/mol)/MW (0.2273 Kg/mol) and $\rho_\pi = 7.7$ Osm/$\text{Kg}_{\text{solute}}$ it would require ~1.30 Kg CHP per Kg water to generate a 10 Osm/Kg solution (~56 wt%). The difference between a 22 wt% solution and 56 wt% can be enormous in terms of rheological properties. These issues are compounded when carrying capacity is also considered. In Figure 6.4 the volumetric carrying capacity (y-axis) is plotted for three solutions with different osmotic densities (various line styles) provided an initial draw solution osmotic pressure (x-axis) and final draw osmotic pressure (various colors).

In Figure 6.4 the low osmotic density solution, X (4.0 Osm/$\text{Kg}_{\text{solute}}$), is limited to an initial concentration of 9 Osm/Kg corresponding to a 75 wt% given that a higher concentration A) is likely a poor candidate rheologically and B) would have diminishing returns with increased concentration. The final draw osmotic pressure (line color) will largely be determined by the feed solution being treated. The initial concentration (maximum available osmotic pressure) and osmotic density determine the volumetric carrying capacity. The impact of carrying capacity cascades through an entire FO water treatment system in terms of the size of equipment and the volumes of intermediate fluids processed to yield a product volume.

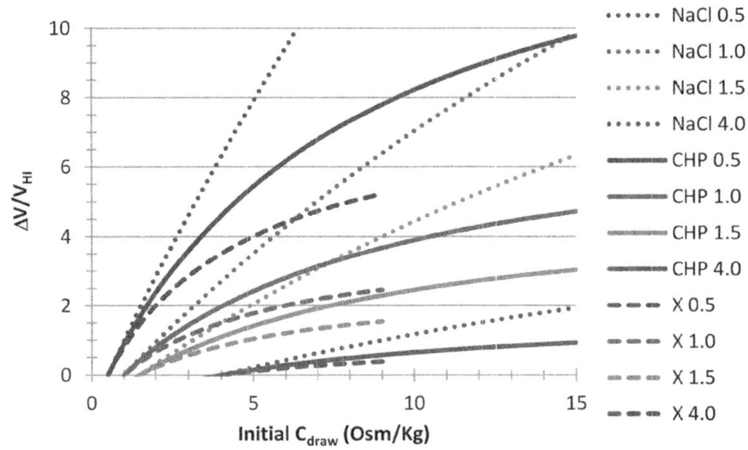

Figure 6.4 The dependence of carrying capacity ($\Delta V/C_{HI}$) on the initial draw concentration (Osm/Kg) is provided for a range of final draw concentrations (black 0.5 Osm/Kg, dark grey 1.0 Osm/Kg, and light grey 1.5 Osm/Kg) and the draw solutes osmotic density (dotted lines NaCl $\rho_{\pi} = 35.8$ Osm/Kg$_{solute}$, solid lines CHP $\rho_{\pi} = 7.6$ Osm/Kg$_{solute}$, and dashed lines X $\rho_{\pi} = 4.0$ Osm/Kg$_{solute}$). Solute X has been limit to 9 Osm/Kg which corresponds to 75 wt%.

6.2.1.7 Osmotic Cost

Assuming all the thermodynamic considerations for a draw solute are well met a rough technoeconomic assessment is warranted. The osmotic density can be converted to an osmotic cost density, ρ_c, based on solutes known or projected material cost (c = currency/mass), Equation 13.

$$\rho_c = \frac{i}{MWc} \tag{13}$$

Even if chemical are not produced at the commodity scale their expected cost can be anticipated. Simple polymers and simple purified organic materials can generally be expected to be between 0.1–1.0 USD/Kg. More complex small molecules and polymers can generally be expected to be between 1.0–10.0 USD/Kg. Exotic small molecules, exotic polymers, and most nano-materials can generally be expected to be great than 10.0 USD/Kg. The high cost of complex materials combined with a low osmotic density can prevent commercial viability.

In our recent study (Reimund *et al.* 2015) we found that a PRO draw solute must have an osmotic cost of less than 9.6 Osm/USD for the solute to match the energy storage cost of a 300 USD/kWh lithium battery. Given that a PRO system has many costs beyond the draw solute and various process inefficiencies it is likely that the osmotic cost would need to be an order of magnitude higher. This is a very challenging cost threshold and very few organic draw solutes will surpass this osmotic cost metric. PRO solutes used for energy storage appear to be limited to simple inorganic salt and perhaps ammonia-CO_2 and trimethylamine-CO_2.

The value of water varies greatly with geography, volume required, and season. In general water had more value than stored electricity at the grid level and a draw solute can be cycled multiple times in a 24 hour period. This means that a FO draw solute will not have the same constraints as PRO draw solute used for energy storage. What constitutes a cost effective draw solute for water treatment must also address the osmotic cost, carrying capacity, rate at which the solute is cycled through the system, and the number times it can be cycled.

6.2.1.8 Solute cycle rate

Once the carrying capacity and osmotic cost are established the only the cycle rate (τ) and solute life time (λ) are required to pieces the life cycle cost of the draw solute per unit of water product water, Equations 14 and 15.

$$\Lambda_{\text{draw}} = \frac{1}{\tau\lambda K_{\Delta V}\rho_c} \tag{14}$$

$$\left[\left(\left(\frac{cycles}{hr}\right)\right)\left(\left(\frac{hrs}{life}\right)\right)\left(\left(\frac{Kg_{\text{water}}}{Osm_{\text{solute}}}\right)\right)\left(\left(\frac{Osm_{\text{solute}}}{USD}\right)\right)\right]^{-1} = \left(\frac{USD}{Kg_{\text{water}}}\frac{life}{cycle}\right) \tag{15}$$

Any hypothetical draw solute life cycle cost can be conducted with dimethylcyclohexylamine draw solute (Stone *et al.* 2013). Dimethylcyclohexylamine draw solute can be estimated at approximately 1 USD/Kg at the ton scale. The cost of CO_2 is ~25 USD/mt (0.001 USD/mol) from a pipe line which is less than 1% the cost of the amines per mole cost of 0.127 USD/mol. Factoring in the van't Hoff index of 1.74 and inverting the cost results in a value of 13.7 Osm/USD. The carrying capacity depends on the process. An initial draw solution concentration of 15 Osm/Kg and a final draw concentration of 2 Osm/Kg would be sufficient to remove most of the water form a seawater feed solution. Using these values with Equation 10 indicates a carrying capacity of 0.43 Kg of water per Osm of draw solute. The life span of the amine involves fugitive losses, trace material lost to water polishing processes, and chemical degradation. The draw will be contaminated overtime and must be regenerated. The amines can be readily purified through standard distillation practices. These losses and regeneration costs can be captured in the amines life time (λ) which for the purposes of this example has been estimated as 5 years ($\lambda = 43{,}800$ hrs/life). The cycle time (τ) through the process involves the FO loop, degasser, mechanical liquid separator, and draw solution regenerating gas contactor. A reasonable estimate for cycle time is 3 hours ($\tau = 0.33$ cycle/hr). These values can be used to solve Equation 14 resulting in a draw solute like cycle cost of 0.012 USD/m^3 of product water. While the cost of the draw solute can be expected to be one of the minor costs for most FO processes it is not negligible. Exotic draw materials with significant cost should be considered critically. Equation 14 provides method to evaluate a draw solute very early in the draw solute development process.

6.2.1.9 Mass Transport

While process kinetics is the ultimate test for an FO process they should only be considered if the thermodynamics are favorable. If the thermodynamics are not favorable the kinetics could be instantaneous and the system would simply waste resources faster. The current theory of internal and external membrane concentration polarization is well document in a wide number of locations. Internal concentration polarization's effect on water flux can be modeled with Equation 16 and variations (Lee *et al.* 1981; Loeb *et al.* 1997).

$$J_w = \frac{1}{K}\ln\left(\frac{A\pi_{\text{draw}} + B}{A\pi_{\text{feed}} + B + J_w}\right) \tag{16}$$

$$K = \frac{t\tau}{\varepsilon D} = \frac{S}{D} \tag{17}$$

This approximation of K, Equation 17, assumes a proportion relationship between the infinite dilution diffusion of a solute in water to its diffusion in a porous support membrane dependent on the porosity and tortuosity of the membrane. The relationships in Equations 16 and 17 are based on derivations of models used to describe RO flux performance. The models have been most widely used for membrane development to quantitatively explain improvements of membranes under development over the existing state-of-the-art commercial membrane. The draw solutes only kinetic parameter in the expression is the draw solutes aqueous diffusion at infinite dilution. This does not take into account diffusion is a concentration dependent parameter whose dependence will differ based on the media (porous support vs. solvent) and differ between various solutes.

It is well known that a high diffusivity is a desirable quality in a feed and draw solution to minimize the effects of concentration polarization and flux kinetics. There is some confusion regarding the chemical properties that govern molecular diffusivity. A few papers have mistakenly correlated diffusivity exclusively with a simple chemical property like solute volume on the molecular scale. This claim is further confused by ambiguous understandings of volume as it relates to monovalent, divalents, monoatomic, polyatomics, cation, anions, and neutrals. Increasing the charge on mono-atomic cation results in a smaller ionic radius as electrons are stripped away (Na^+ 116 pm vs. Mg^{2+} 86 pm), however the charge density increases which increases the number of "waters of hydration" increasing the ion's effective mass, volume, and hydration radius (Na^+ 355 pm vs. Mg^{2+} 410 pm). Furthermore smaller an ions have greater the charge density and thus the greater hydration for example Li^+ has ~6 water of hydration vs. Na^+ ~4 waters of hydration (Wilson & Stewart, 2013; Zavitsas, 2001). This trend partial holds for anions with F^- diffusing significantly slower than the other halides however the remaining halides are very similar due their limited hydration. Diffusion of similar organic molecules and polymers are polymers are predominantly based on the molecular volume. However chemically different organics will have different interaction between the solvent and other solutes resulting in significantly different diffusivities and viscosities. Mixed system such as metal salts of polycarboxylic acids can also feature components with very different diffusion rates.

In generally draw solutes with larger molecular volumes will have lower diffusion rate and higher viscosities the external polarization plays a larger role. It was found with using different concentrations of DMCHA that solute derived external polarization was more significant than membrane polarization (Reimund *et al.* 2016) and the rheological property differences between DMCHA and CHP, Figure 6.5, under identical conditions, Figure 6.5, CHP has 23% less flux (11 vs. 8.5 L/m^{2}hr) (Orme & Wilson, 2015). There are few well designed studies which quantitatively measured the effect that mass transport phenomenon influence flux rates. Looking broadly at the literature it is clear that high molecular mass, low diffusion solutes, and high viscosity solutions result in lower flux rates for similar osmotic concentrations (Fang *et al.* 2013; Zhao & Zou, 2011; Gray *et al.* 2006). Matsuyama is the first to look at this issue systematically (Yasukawa *et al.* 2015) using a range of PEG oligomers.

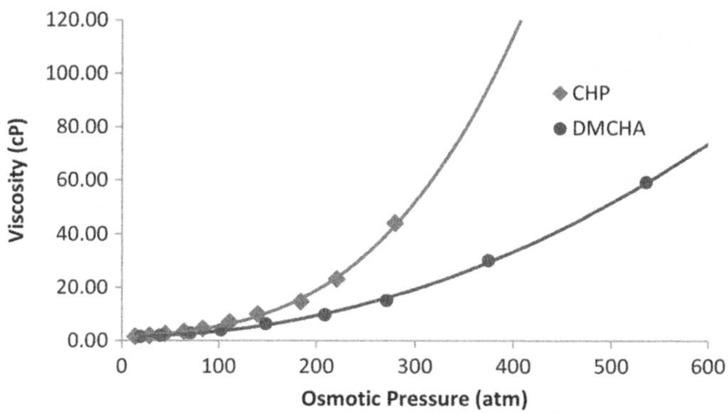

Figure 6.5 The absolute viscosity (cP) recorded at 15°C plotted against osmotic pressure of the polar bicarbonate solutions of CHP (blue diamonds, 186.3 g/mol) and DMCHA (red circles, 127.2 g/mol). CHP and DMCHA data extends beyond depicted osmotic pressure and viscosity range. The trend lines are based on 3rd order polynomials.

With this said low viscosity is not intrinsically required for an ODMP, its possible for a stationary phase (Razmjou *et al.* 2013) or near stationary phases to change the potential of water at their surface. Capillary action and swelling is when a solid phase adjusts the solution structure or potential of water, which is directly related to osmotic pressure where a solute adjusts the solution structure or potential of water. At the molecular scale the differences between capillary action and osmotic pressure phenomena are minor. The energies of swelling pressures can be an order magnitude larger than osmotic pressure (Refojo, 1976). Despite the theoretical viability, highly viscous draw solutions generally have mass transfer issues that reduce pure water flux during an FO process. These mass transport issues include difficulty in achieving proper draw solution mixing due to shear thickening or thinning, which are non-ideal behaviors which generally increase with viscosity.

6.2.1.10 Membrane Permeability – Reverse Solute Flux

Membrane permeability of draw solutes through FO membranes largely depends on a solutes molecular volume, hydration sphere, and a variety membrane properties. Conventional draw solutes like alkali and alkaline halide salts, ammonia-CO_2, and polar organics like ethanol tend to have noticeable membrane permeability. This membrane permeability is expressed as a solute flux (J_s) but is more meaningful when expressed as a ratio of the water flux (J_w) to the solute flux to (J_w/J_s) also known as the reverse flux selectivity. The reverse flux selectivity is dependent on the membrane and draw solute combination. Factors such as solute/membrane solubility, solute/membrane charge/polarity, molecular volume/mass factor into the reverse flux selectivity. The trend towards high molecular mass draw solutes is producing draw solutes with much lower reverse solute flux rendering it a a less significant concern for many of the system under consideration.

There has been at least one paper discussing concern around reactive reverse solute flux. This is when a component of the draw solute can react with a component of the feed solution. This is a specific concern for carbonate based draw systems. If the feed stream contains a cation whose carbonate complex has a very low solubility reactive reverse solute flux could induce scaling at the membrane surface or solid formation in the membrane.

6.2.2 Types of draw solute

An ODMP draw solute can largely be categorized by how it is regenerated (or used) once diluted. In the text that follows a few notable examples of the draw solute type are highlighted. The primary goal is to offer a broad evaluation of the separation method in terms of thermodynamics based techno-economics.

6.2.2.1 Osmotic filtration

Osmotic filtration is a process that removes a dissolved solid. A number of the other draw solute regeneration processes remove the bulk of the water through a stimulus driven process than finish the process with an osmotic filtration. Ignoring charge and hydration attenuations a small molecular volume dissolved solids, such as NaCl, require a small pore membrane, such a RO membrane, to achieve good separations in a filtration process. As molecular volumes increase dissolved solids (again ignoring charge and hydration considerations) allows the use of membranes with increased pore size such as nanofiltration or ultrafiltration membranes. However, the energy required to concentrate a dissolved solid is not based on the membrane porosity. The thermodynamic energy necessary to dewater dissolved solute is independent of the method employed and membrane pore size.

Ultrafiltration membranes (and NF membranes) are usually used to remove suspended materials which have no appreciable osmotic energy cost. Again, regeneration of draw solutes, which are dissolved solids, will have the same osmotic cost for an RO membrane and larger pore ultrafiltration membranes. The wide spread understanding of ultrafiltration energy costs when applied to suspended solids, its traditional use, has led to confusion as ultrafiltration has been reapplied to dissolved solids. Regeneration of a draw solute with a large-pore membrane is not fundamentally a lower-pressure lower-energy process than RO. This misconception has been intentionally or unintentionally encouraged by the way osmotic filtration data has been presented in the literature. There are examples in the literature where research groups conduct their FO test at concentration between 0.5–2.0 M with osmotic pressure between 20–100 atm then "demonstrate" the osmotic filtration regeneration with solutions that never rise above 0.1 M. Using low concentration solutions allow low hydraulic pressure (10 atm) osmotic filtration but this "demonstrates" only partial regenerations. To fully regenerate the solute it would require an osmotic filtration process with hydraulic process equivalent to the FO draw solutes highest osmotic pressure in the process. Thus the osmotic filtration process sets the maximum osmotic draw solute pressure. Because of concentration polarization effects the osmotic filtration regeneration steps is limited to the same pressures as conventional RO (~50 atm). FO systems which are regenerated exclusively with osmotic filtration are essentially pretreatments for processes that have the same concentration limits as conventional RO.

The free energy mixing (ΔG_{mix}) is often used to approximate energy required to dewater a solution. The free energy method uses many values that assume infinite dilution and approximations that are only valid under dilute conditions. With a reliable relationship between a solutions osmotic pressure and concentration (Wilson & Stewart, 2013) the minimum energy required to remove and add pure water energy can easily be calculated in the form of L•bar, from Equation 18. The only values required to make this calculation is the solutions initial osmotic pressure (π_i), initial solution water fraction (w_i), and fraction of water removed or added (ΔV) (Reimund *et al.* 2015).

$$W = \int_0^{\Delta V} (\pi_i V_i w_i \ln(V_i w_i + \Delta V)) d\Delta V \tag{18}$$

Advantages of using more porous membrane with large molecular volume draws over RO include the potentially for reduced membrane cost, reduced pressure drop, and higher water fluxes. These advantages must counter balance cost associated with solute osmotic density, mass transport issues, solute osmotic cost, and limits on the concentrations of solutions that can be treated when compared to other FO processes.

Draw solutes that have been explored for osmotic filtration include molecular polycarboxylates (Ge *et al.* 2012; Hau *et al.* n.d.; Stone *et al.* 2013; Zhao *et al.* 2014), inorganic complexes (Ge & Chung, 2013; Ge *et al.* 2014), quantum dots (Guo *et al.* 2014), and micellar systems (Gadelha *et al.* 2014; Roach *et al.* 2014).

6.2.2.2 Membrane distillation

Membrane distillation (MD) is used in a number ODMP draw solute regeneration systems specifically in the PRO processes. Organic salts and molecules based on 2-methylimidazole have been considered for MD re-concentration (Yen *et al.* 2010). The advantage of MD is the production of high purity product from highly concentrated solutions using low-grade heat. This allows a MD regenerated draw to reach much higher concentrations than a draw regenerated with osmotic filtration (UF, NF, and RO). The disadvantage of MD is the large amount of low-grade heat and capital cost of the infrastructure required to deliver that heat. The gained output ratio (GOR) is the ratio of the latent heat of evaporation in unit water to the amount of energy used by a desalination system to produce a unit of product water. The latent heat of evaporation is much higher than the minimum energy of water purification and thus much of the energy can be recycled when the product water is re-condensed. Highly efficient distillation processes can recycle the latent heat of evaporation and have high GORs up to 12. MD systems are not especially efficient with GORs between 0.5 to 4 and thus require "waste" heat to be cost effective. MD as all membrane processes is prone to scaling and fouling; FO is expected to be resistant to fouling and offers MD with a consistent homogenous feed stream.

6.2.2.3 Unremoved draw solutes

Strictly speaking these processes do no generate water; they generate useful solution from impaired water sources. Hydration Technology Invitations (HTI) had a number of products that operate by using a sugar electrolyte mixture as a draw solute to produce a consumable solution from a contaminated or saline feed source. There have also been a series of papers discussing the use of ODMP for "fertigation" in which fertilizer is used as the draw solute to obtain water from an impaired source. Fertigation produces a solution which can both irrigate and fertilize a field (Phuntsho *et al.* 2012; Phuntsho *et al.* 2012; Phuntsho *et al.* 2011).

6.2.2.4 Magnetic draw solute

Magnetic draw solutes are materials with appreciable osmotic pressure which can be isolated magnetically (Ge *et al.* 2010; Ling *et al.* 2010; Zhao *et al.* 2013). Presumably the magnetic energy will be manipulated electronically. Given the efficiency of delivering and recycling hydraulic pressure used in osmotic filtration it is unlikely that a magnetic manipulation could reduce the energy requirements of water treatment. Even if the energy requirements of regeneration could be made competitive, the molecular masses of nanoparticles are substantial. This size generally reduces their maximum osmotic pressure, reduces osmotic density, reduces carrying capacity, increases the osmotic cost, and restricts mass transport properties. In terms of practical concern magnetic draw need to be successfully regenerated. Even if magnetic draw solutes fail the commercialization test for water treatment there is still a number of reasons to study the osmotic activity of magnetic materials. ODMP is more than just water treatment. Linking the osmotic pressure of a solution to a magnetic field at the nanoscale may be a key step in designing the nanosystems.

6.2.2.5 Stoichiometric chemically reactive

A few systems have been proposed that utilize salt metathesis or a similar mechanism to remove a draw solute in a FO process (Frank, 1972; Alnaizy *et al.* 2013a, 2013b). In these systems the draw solute is soluble with good osmotic concentration. Once diluted in an ODMP process the water can be obtained through the addition of a stoichiometric amount of a second material which converts the draw solute to an insoluble form. For example a diluted magnesium sulfate solution could be reacted with barium hydroxide to form insoluble magnesium hydroxide and insoluble barium sulfate, Equation 19 (Alnaizy *et al.* 2013b). The magnesium sulfate can be regenerated through the addition of sulfuric acid to the magnesium hydroxide but the barium sulfate is a "byproduct" and most likely a waste.

$$MgSO_{4(aq)} + Ba(OH)_{2(aq)} \rightarrow Mg(OH)_{2(s)} + BaSO_{4(s)} \tag{19}$$

These salt metathesis ODMP systems have no fundamental process energy requirements; both the FO process and formation of insoluble form are spontaneous processes. This allows for very limited energy use at the point of water treatment but requires precipitating agent and either the draw solute or regenerating agent to be delivered to the point of use and insoluble product removed. Enlarging the energy evaluation to include the cost of generating or regenerating the precipitating agent must be consider to compare salt metathesis ODMP systems to other ODMP systems. Once this is done it is clear that such systems are rarely economically competitive.

Salt metathesis ODMP need to need for chemicals delivered point of treatment is similar to many RO pretreatment processes such as ion exchange or precipitory water softening. The needs of salt metathesis ODMP are larger. For reported systems the quantity of materials delivered to water treatment point of operation would be twice the mole quantity (precipitant and regeneration agent) of the salt removed from the product water. It would dramatically improve salt metathesis ODMP processes if A) the stoichiometric ratio of the precipitant and regeneration agent could be reduced or B) if the materials could be effectively regenerated on site.

6.2.2.6 Volatile solutes

There are a number of volatile solutes which have been suggested as draw solutes. In terms of simplified thermodynamics technoeconomic analysis volatile draw solute systems often make sense. The mass transfer kinetics of volatile draw solutes are in many cases favorable for ODMP processes. Exact issues vary with the specific system. The ammonia-CO_2 and trimethylamine-CO_2 (Boo *et al.* 2015) systems are a combination volatile solute and thermolytic solutes. Sulfur dioxide has been considered as a volatile draw solute (George W. Batchelder, 1965; David N. Glew, 1965; McGinnis, 2002) but in overall there are a few exclusively volatile solute draw systems that have been considered and very few of these systems have been followed up.

The challenges with volatile solutes include reverse solute flux and full removal from the product water discussed above as the MSDOC. The ammonia-CO_2 system suffers from both these issues. Ammonia readily passes through the FO membrane into the concentrate. Ultimately the concentrate solution and product water both need to be taken to near water reflux temperatures to remove ammonia (Hancock, 2013).

6.2.2.7 Thermally driven phase change solutes

Thermally driven phase change solutes can be divided into two categories, materials that phase separate from water at 1) high temperatures and 2) low temperatures. Sandia National Laboratory serious explored materials that would precipitate at low temperatures with a large portion of their interest directed to Na_2HPO_4 (Miller & Evans, 2006). Their system required the solution draw solution to be cooled and draw solute precipitated. Their technoeconomic analysis was not favorable and research in the area was concluded circa 2006.

Lower solution critical temperature (LSCT) draw solutions have been published in both the peer review and patent literature (Cai *et al.* 2013; Li *et al.* 2011; Li *et al.* 2013; Li *et al.* 2011; Mok *et al.* 2013; Nakayama *et al.* 2014; Noh *et al.* 2012; Ou *et al.* 2013; Razmjou *et al.* 2013). LSCT material phase separate as the temperature increases; this sort of behavior is usually observed for polymers/ oligomers however there are some small molecules such as triethylamine that display this behavior. Both rich and lean phases in LCST systems generally include moderate fractions of water and solute. This leads to issues with maximum osmotic concentration and MSDOC.

6.2.2.8 *Solid draw agents*

Solid draw agents include both solution suspended and membrane affixed arrangements. The transition between dissolved solid and suspended solid is not always clear, hydrogel polymers often straddle the boundary (Li *et al.* 2011; Li *et al.* 2013; Razmjou *et al.* 2013; Razmjou *et al.* 2013; Li & Wang, 2013). The solid draw agents reported thus far rely on a thermally driven LSCT behavior to obtain the product water.

6.2.2.9 *Thermolytic solutes*

Thermal driven phase changes must rely on temperature dependent physical properties. Thermolytic solutes take this process a step further by using temperature dependent chemical reactions that induce radical changes in the materials physical properties. The result is a less graduated and more quantum transition in solute osmotic characteristics. Thermolytic solutes include the ammonia-CO_2 system invigorated ODMP research as well as a number of additional CO_2 triggered amine systems. Our group has worked with switchable polarity solvents (Stone *et al.* 2013; Wilson & Stewart, 2014; Wilson & Orme, 2015; McNally *et al.* 2015; Orme & Wilson, 2015; Wendt *et al.* 2015), Figure 6.6.

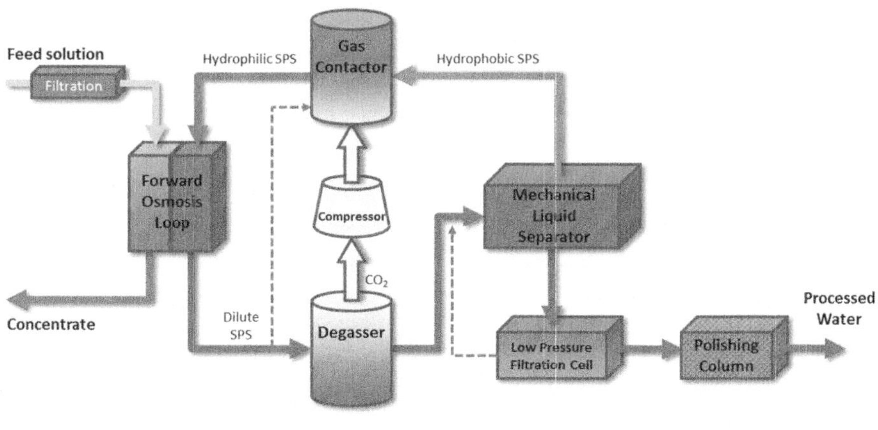

Figure 6.6 The proposed switchable polarity solvent forward osmosis (SPS FO) process.

6.3 CONCLUSION

It is not difficult to find a draw solute that can demonstrate water flux in an ODMP process all it takes is water soluble material. The challenge is finding or designing a draw solute that meets the highly constrained energy, capital, and process cost requirement while supplying good kinetics. For water treatment it is likely the ideal draw will provide high osmotic pressures in its concentrated form and the product water will be removed with low-grade heat making the process more cost effective than electrically driven processes. The high osmotic pressure implies a low molecular mass draw solute with a high osmotic density. With a great deal of research is directed to high molecular mass draw solutes, including nanomaterials, researchers must be extra vigilant in evaluating the relevant fundamental thermodynamic properties are good. It is expected that there unimagined applications of ODMP where low cost, high osmotic pressure and thermal regeneration are not required.

6.4 NOMENCLATURE

A = Pure water permeability, m s^{-1} atm^{-1}
a = activity
B = Salt permeability constant, m s^{-1}
c = cost
C_{sub} = concentration
$C_{aw} = -\ln(a_w/V_m)$ activity concentration proxy
C_h = molality corrected for waters of hydration
$C_{xw} = -\ln(x_w/V_m)$ mole faction concentration proxy
CHP = *N*-cyclohexylpiperidine
D = diffusivity
DMCHA = *N,N*-dimethylcyclohexylamine
FO = forward osmosis
h = total waters of hydration per mole solute
i = van't Hoff index
i_m = mas ratio van't Hoff index
J_w = water flux
J_s = reverse draw solute flux
K = solute resistivity
K_F = cryoscopic constant
$K_{\Delta V}$ = mole ratio carrying capacity
mol = moles
mol/Kg = molality
M = mass
MSDOC = minimum stimuli-driven osmotic concentration
MW = molecular weight
n = number of moles
NF = nanofiltration

Osm/Kg = osmolality
P_{op} = system operation pressure
R − ideal gas constant
RO = reverse osmosis
S = structural parameter
t = thickness
T = absolute temperature
ΔT_F = deviation in the freezing point
UF = ultrafiltration
V_m = molecular volume
V_{sub} = solution volume
ΔV = Water transfer from feed to draw
$\Delta V/V_{HI}$ = volumetric carrying capacity
w_{sub} = weight fraction
wt% = weight percent
x = mole fraction
y = activity coefficient

Greek symbols
Π, π_{sub} = osmotic pressure
ε = porosity
Λ_{sub} = life cycle cost
λ = lifetime or
τ = cycle time or tortuosity
ρ = mass density
ρ_π = osmotic density
ρ_c = osmotic cost density
μ = chemical potential

Subscripts
HI = draw solution input
HO = draw solution output
i = feed for desalination, draw for PRO
LI = feed solution input
LO = feed solution output
p = pressure
s = solute
w = water
x = mole fraction

6.5 REFERENCES

Achilli A. and Childress A. E. (2010). Pressure retarded osmosis: From the vision of Sidney Loeb to the first prototype installation – Review. *Desalination*, **261**(3), 205–211. http://doi.org/10.1016/j.desal.2010.06.017

Achilli A., Cath T. Y. and Childress A. E. (2010). Selection of inorganic-based draw solutions for forward osmosis applications. *Journal of Membrane Science*, **364**(1–2), 233–241. http://doi.org/10.1016/j.memsci.2010.08.010

Alnaizy R., Aidan A. and Qasim M. (2013a). Copper sulfate as draw solute in forward osmosis desalination. *Journal of Environmental Chemical Engineering*, **1**(3), 424–430. http://doi.org/10.1016/j.jece.2013.06.005

Alnaizy R., Aidan A. and Qasim M. (2013b). Draw solute recovery by metathesis precipitation in forward osmosis desalination. *Desalination and Water Treatment*, **51**(28–30), 5516–5525. http://doi.org/10.1080/19443994.2013.770238

Boo C., Khalil Y. F. and Elimelech M. (2015). Performance evaluation of trimethylamine–carbon dioxide thermolytic draw solution for engineered osmosis. *Journal of Membrane Science*, **473**, 302–309. http://doi.org/10.1016/j.memsci.2014.09.026

Cai Y., Shen W., Loo S. L., Krantz W. B., Wang R., Fane A. G. and Hu X. (2013). Towards temperature driven forward osmosis desalination using Semi-IPN hydrogels as reversible draw agents. *Water Research*, **47**(11), 3773–3781. http://doi.org/10.1016/j.watres.2013.04.034

Cath T. Y., Childress A. E. and Elimelech M. (2006). Forward osmosis: Principles, applications, and recent developments. *Journal of Membrane Science*, **281**(1–2), 70–87.

Chekli L., Phuntsho S., Shon H. K., Vigneswaran S., Kandasamy J. and Chanan A. (2012). A review of draw solutes in forward osmosis process and their use in modern applications. *Desalination and Water Treatment*, **43**(1–3), 167–184. http://doi.org/10.1080/19443994.2012.672168

Chung T.-S., Zhang S., Wang K. Y., Su J. and Ling M. M. (2012). Forward osmosis processes: Yesterday, today and tomorrow. *Desalination*, **287**, 78–81. http://doi.org/10.1016/j.desal.2010.12.019

David N. G. (1965, November 9). United States Patent: 3216930 – Process for Liquid Recovery and Solution Concentration. Retrieved from http://patft.uspto.gov/netacgi/nph-Parser?Sect2=PTO1&Sect2=HITOFF&p=1&u=%2Fnetahtml%2FPTO%2Fsearch-bool.html&r=1&f=G&l=50&d=PALL&RefSrch=yes&Query=PN%2F3216930

Fang Y., Bian L. and Wang X. (2013). Understanding membrane parameters of a forward osmosis membrane based on nonequilibrium thermodynamics. *Journal of Membrane Science*, **437**, 72–81. http://doi.org/10.1016/j.memsci.2013.02.052

Frank B. S. (1972, June 20). United States Patent: 3670897 – Desalination of Sea Water. Retrieved from http://patft.uspto.gov/netacgi/nph-Parser?Sect2=PTO1&Sect2=HITOFF&p=1&u=%2Fnetahtml%2FPTO%2Fsearch-bool.html&r=1&f=G&l=50&d=PALL&RefSrch=yes&Query=PN%2F3670897

Gadelha G., Nawaz M. S., Hankins N. P., Khan S. J., Wang R. and Tang C. Y. (2014). Assessment of micellar solutions as draw solutions for forward osmosis. *Desalination*, **354**, 97–106. http://doi.org/10.1016/j.desal.2014.09.009

Ge Q. and Chung T.-S. (2013). Hydroacid complexes: a new class of draw solutes to promote forward osmosis (FO) processes. *Chemical Communications*, **49**(76), 8471–8473. http://doi.org/10.1039/C3CC43951H

Ge Q., Su J., Chung T.-S. and Amy G. (2010). Hydrophilic Superparamagnetic Nanoparticles: Synthesis, Characterization, and Performance in Forward Osmosis Processes. *Ind. Eng. Chem. Res.*, **50**(1), 382–388. http://doi.org/10.1021/ie101013w

Ge Q., Su J., Amy G. L. and Chung T.-S. (2012). Exploration of polyelectrolytes as draw solutes in forward osmosis processes. *Water Research*, **46**(4), 1318–1326. http://doi.org/10.1016/j.watres.2011.12.043

Ge Q., Ling M. and Chung T.-S. (2013). Draw solutions for forward osmosis processes: Developments, challenges, and prospects for the future. *Journal of Membrane Science*, **442**, 225–237. http://doi.org/10.1016/j.memsci.2013.03.046

Ge Q., Fu F. and Chung T.-S. (2014). Ferric and cobaltous hydroacid complexes for forward osmosis (FO) processes. *Water Research*, **58**, 230–238. http://doi.org/10.1016/j.watres.2014.03.024

George W. B. (1965, March 2). United States Patent: 3171799 – Process for the demineralization of Water. Retrieved from http://patft.uspto.gov/netacgi/nph-Parser?Sect2=PTO1&Sect2=HITOFF&p=1&u=%2Fnetahtml%2FPTO%2Fsearch-bool.html&r=1&f=G&l=50&d=PALL&RefSrch=yes&Query=PN%2F3171799

Gray G. T., McCutcheon J. R. and Elimelech M. (2006). Internal concentration polarization in forward osmosis: role of membrane orientation. *Desalination*, **197**(1–3), 1–8. http://doi.org/10.1016/j.desal.2006.02.003

Guo C. X., Zhao D., Zhao Q., Wang P. and Lu X. (2014). Na + -functionalized carbon quantum dots: a new draw solute in forward osmosis for seawater desalination. *Chemical Communications*, **50**(55), 7318–7321. http://doi.org/10.1039/C4CC01603C

Hancock N. (2013). Engineered Osmosis for Energy Efficient Separations: Optimizing Waste Heat Utilization FINAL SCIENTIFIC REPORT DOE F 241.3 DE-EE0003467. Retrieved from http://www.osti.gov/bridge/servlets/purl/1067379/

Hau N. T., Chen S.-S., Nguyen N. C., Huang K. Z., Ngo H. H. and Guo W. (n.d.). Exploration of EDTA sodium salt as Novel Draw Solution in Forward Osmosis Process for Dewatering of High Nutrient Sludge. *Journal of Membrane Science*. http://doi.org/10.1016/j.memsci.2013.12.068

Hoover L. A., Phillip W. A., Tiraferri A., Yip N. Y. and Elimelech M. (2011). Forward with Osmosis: emerging applications for greater sustainability. *Environmental Science & Technology*, **45**(23), 9824–9830. http://doi.org/10.1021/es202576h

James E. M. and Lindsey R. E. (2006). SAND2006-4634 Forward Osmosis: A New Approach to Water Purification and Desalination (SANDIA REPORT No. SAND2006-4634). Sandia National Laboratory.

Lee K. L., Baker R. W. and Lonsdale H. K. (1981). Membranes for power generation by pressure-retarded osmosis. *Journal of Membrane Science*, **8**(2), 141–171. http://doi.org/10.1016/S0376-7388(00)82088-8

Li D. and Wang H. (2013). Smart draw agents for emerging forward osmosis application. *Journal of Materials Chemistry A*, **1**(45), 14049–14060. http://doi.org/10.1039/C3TA12559A

Li D., Zhang X., Yao J., Simon G. P. and Wang H. (2011). Stimuli-responsive polymer hydrogels as a new class of draw agent for forward osmosis desalination. *Chemical Communications*, **47**(6), 1710. http://doi.org/10.1039/c0cc04701e

Li D., Zhang X., Yao J., Zeng Y., Simon G. P. and Wang H. (2011). Composite polymer hydrogels as draw agents in forward osmosis and solar dewatering. *Soft Matter*, **7**(21), 10048–10056. http://doi.org/10.1039/C1SM06043K

Li D., Zhang X., Simon G. P. and Wang H. (2013). Forward osmosis desalination using polymer hydrogels as a draw agent: influence of draw agent, feed solution and membrane on process performance. *Water Research*, **47**(1), 209–215. http://doi.org/10.1016/j.watres.2012.09.049

Ling M. M., Wang K. Y. and Chung T.-S. (2010). Highly water-soluble magnetic nanoparticles as novel draw solutes in forward osmosis for water reuse. *Ind. Eng. Chem. Res.*, **49**(12), 5869–5876. http://doi.org/10.1021/ie100438x

Loeb S., Titelman L., Korngold E. and Freiman J. (1997). Effect of porous support fabric on osmosis through a Loeb-Sourirajan type asymmetric membrane. *Journal of Membrane Science*, **129**(2), 243–249. http://doi.org/10.1016/S0376-7388(96)00354-7

McCutcheon J. R., McGinnis R. L. and Elimelech M. (2005). A novel ammonia–carbon dioxide forward (direct) osmosis desalination process. *Desalination*, **174**(1), 1–11. http://doi.org/10.1016/j.desal.2004.11.002

McGinnis R. L. (2002, May 21). United States Patent: 6391205 – Osmotic desalinization process. Retrieved from http://www.google.com/patents?id=umIKAAAAEBAJ

McNally J. S., Noll B., Orme C. J. and Wilson A. D. (2015). Density functional theory analysis of the impact of steric interaction on the function of switchable polarity solvents. *The Journal of Physical Chemistry B*, **119**(22), 6766–6775. http://doi.org/10.1021/acs.jpcb.5b03167

Mok Y., Nakayama D., Noh M., Jang S., Kim T. and Lee Y. (2013). Circulatory osmotic desalination driven by a mild temperature gradient based on lower critical solution temperature (LCST) phase transition materials. *Physical Chemistry Chemical Physics*, **15**(44), 19510–19517. http://doi.org/10.1039/C3CP52613E

Nakayama D., Mok Y., Noh M., Park J., Kang S. and Lee Y. (2014). Lower critical solution temperature (LCST) phase separation of glycol ethers for forward osmotic control. *Physical Chemistry Chemical Physics*, **16**(11), 5319–5325. http://doi.org/10.1039/C3CP55467H

Noh M., Mok Y., Lee S., Kim H., Lee S. H., Jin G., … Lee Y. (2012). Novel lower critical solution temperature phase transition materials effectively control osmosis by mild temperature changes. *Chem. Commun.*, **48**(32), 3845–3847. http://doi.org/10.1039/C2CC30890H

Orme C. J. and Wilson A. D. (2015). 1-Cyclohexylpiperidine as a thermolytic draw solute for osmotically driven membrane processes. *Desalination*, **371**, 126–133. http://doi.org/10.1016/j.desal.2015.05.024

Ou R., Wang Y., Wang H. and Xu T. (2013). Thermo-sensitive polyelectrolytes as draw solutions in forward osmosis process. *Desalination*, **318**, 48–55. http://doi.org/10.1016/j.desal.2013.03.022

Pattle R. E. (1954). Production of electric power by mixing fresh and salt water in the hydroelectric pile. *Nature*, **174**(4431), 660–660. http://doi.org/10.1038/174660a0

Phuntsho S., Shon H. K., Hong S., Lee S. and Vigneswaran S. (2011). A novel low energy fertilizer driven forward osmosis desalination for direct fertigation: evaluating the performance of fertilizer draw solutions. *Journal of Membrane Science*, **375**(1–2), 172–181. http://doi.org/10.1016/j.memsci.2011.03.038

Phuntsho S., Shon H. K., Hong S., Lee S., Vigneswaran S. and Kandasamy J. (2012). Fertiliser drawn forward osmosis desalination: the concept, performance and limitations for fertigation. *Reviews in Environmental Science and Bio/Technology*, **11**(2), 147–168. http://doi.org/10.1007/s11157-011-9259-2

Phuntsho S., Shon H. K., Majeed T., Saliby I. El, Vigneswaran S., Kandasamy J., … Lee S. (2012). Blended fertilizers as draw solutions for fertilizer-drawn forward osmosis desalination. *Environmental Science & Technology*, **46**(8), 4567–4575. http://doi.org/10.1021/es300002w

Qasim M., Darwish N. A., Sarp S. and Hilal N. (2015). Water desalination by forward (direct) osmosis phenomenon: a comprehensive review. *Desalination*, **374**, 47–69. http://doi.org/10.1016/j.desal.2015.07.016

Ray A. N. (1964, April 21). United States Patent: 3130156 – Solvent Extractor. Retrieved from http://patft.uspto.gov/netacgi/nph-Parser?Sect2=PTO1&Sect2=HITOFF&p=1&u=%2Fnetahtml%2FPTO%2Fsearch-bool.html&r=1&f=G&l=50&d=PALL&RefSrch=yes&Query=PN%2F3130156

Razmjou A., Liu Q., Simon G. P. and Wang H. (2013). Bifunctional polymer hydrogel layers as forward osmosis draw agents for continuous production of fresh water using solar energy. *Environmental Science & Technology*, **47**(22), 13160–13166. http://doi.org/10.1021/es403266y

Razmjou A., Simon G. P. and Wang H. (2013). Effect of particle size on the performance of forward osmosis desalination by stimuli-responsive polymer hydrogels as a draw agent. *Chemical Engineering Journal*, **215–216**, 913–920. http://doi.org/10.1016/j.cej.2012.11.088

Refojo M. F. (1976). Vapor pressure and swelling pressure of hydrogels. In: Hydrogels for Medical and Related Applications, J. D. Andrade (ed.), Vol. **31**. Washington D. C. American Chemical Society. pp. 37–51. Retrieved from http://pubs.acs.org/doi/abs/10.1021/bk-1976-0031.ch002

Reimund K. K., Coscia B. J., Arena J. T., Wilson A. D. and McCutcheon J. R. (2016). Characterization and membrane stability study for the switchable polarity solvent N,N-Dimethylcyclohexylamine as a draw solute in forward osmosis. *Journal of Membrane Science*, **501**, 93–99. http://doi.org/10.1016/j.memsci.2015.10.039.

Reimund K. K., McCutcheon J. R. and Wilson A. D. (2015). Thermodynamic analysis of energy density in pressure retarded osmosis: The impact of solution volumes and costs. *Journal of Membrane Science*, **487**, 240–248. http://doi.org/10.1016/j.memsci.2015.03.076

Roach J., Al-Abdulmalek A., Al-Naama A. and Haji M. (2014). Use of micellar solutions as draw agents in forward osmosis. *Journal of Surfactants and Detergents*, **17**(6), 1241–1248. http://doi.org/10.1007/s11743-014-1638-6

Schaack J. R. H. V. (1932, November 1). Process of drying liquids by dialysis. Retrieved from http://www.google.com/patents/US1885393

Shaffer D. L., Werber J. R., Jaramillo H., Lin S. and Elimelech M. (2015). Forward osmosis: Where are we now? *Desalination*, **356**, 271–284. http://doi.org/10.1016/j.desal.2014.10.031

Stone M. L., Rae C., Stewart F. F. and Wilson A. D. (2013). Switchable polarity solvents as draw solutes for forward osmosis. *Desalination*, **312**, 124–129. http://doi.org/10.1016/j.desal.2012.07.034

Stone M. L., Wilson A. D., Harrup M. K. and Stewart F. F. (2013). An initial study of hexavalent phosphazene salts as draw solutes in forward osmosis. *Desalination*, **312**, 130–136. http://doi.org/10.1016/j.desal.2012.09.030

van't Hoff J. H. (1995). The role of osmotic pressure in the analogy between solutions and gases. *Journal of Membrane Science*, **100**(1), 39–44.

Wang H., Wei J. and Simon G. P. (2014). Response to osmotic pressure versus swelling pressure: comment on "Bifunctional Polymer Hydrogel Layers As Forward Osmosis Draw Agents for Continuous Production of Fresh Water Using Solar Energy." *Environmental Science & Technology*, **48**(7), 4214–4215. http://doi.org/10.1021/es5011016

Wendt D. S., Orme C. J., Mines G. L. and Wilson A. D. (2015). Energy requirements of the switchable polarity solvent forward osmosis (SPS-FO) water purification process. *Desalination*, **374**, 81–91. http://doi.org/10.1016/j.desal.2015.07.012

Wilson A. D. and Orme C. J. (2015). Concentration dependent speciation and mass transport properties of switchable polarity solvents. *RSC Advances*, **5**(10), 7740–7751. http://doi.org/10.1039/C4RA08558B

Wilson A. D. and Stewart F. F. (2013). Deriving osmotic pressures of draw solutes used in osmotically driven membrane processes. *Journal of Membrane Science*, **431**, 205–211. http://doi.org/10.1016/j.memsci.2012.12.042

Wilson A. D. and Stewart F. F. (2014). Structure-function study of tertiary amines as switchable polarity solvents. *RSC Advances*, **4**(22), 11039–11049. http://doi.org/10.1039/C3RA47724J

Yasukawa M., Tanaka Y., Takahashi T., Shibuya M., Mishima S. and Matsuyama H. (2015). Effect of molecular weight of draw solute on water permeation in forward osmosis process. *Industrial & Engineering Chemistry Research*. http://doi.org/10.1021/acs.iecr.5b01960

Yen S. K., Mehnas Haja N. F., Su M., Wang K. Y. and Chung T.-S. (2010). Study of draw solutes using 2-methylimidazole-based compounds in forward osmosis. *Journal of Membrane Science*, **364**(1–2), 242–252. http://doi.org/10.1016/j.memsci.2010.08.021

Zavitsas A. A. (2001). Properties of water solutions of electrolytes and nonelectrolytes. *The Journal of Physical Chemistry B*, **105**(32), 7805–7817. http://doi.org/10.1021/jp011053l

Zhao D., Chen S., Wang P., Zhao Q. and Lu X. (2014). A Dendrimer-based forward osmosis draw solute for seawater desalination. *Industrial & Engineering Chemistry Research*, **53**(42), 16170–16175. http://doi.org/10.1021/ie5031997

Zhao S. (2014). Osmotic pressure versus swelling pressure: comment on "Bifunctional Polymer Hydrogel Layers As Forward Osmosis Draw Agents for Continuous Production of Fresh Water Using Solar Energy." *Environmental Science & Technology*, **48**(7), 4212–4213. http://doi.org/10.1021/es5006994

Zhao S. and Zou L. (2011). Relating solution physicochemical properties to internal concentration polarization in forward osmosis. *Journal of Membrane Science*, **379**(1–2), 459–467. http://doi.org/10.1016/j.memsci.2011.06.021

Zhao Q., Chen N., Zhao D. and Lu X. (2013). Thermoresponsive magnetic nanoparticles for seawater desalination. *ACS Applied Materials & Interfaces*, **5**(21), 11453–11461. http://doi.org/10.1021/am403719s

Chapter 7

Nanotechnology for microbial fuel cells

Muhammad Mustafa Hussain

With increased global population, need for basic lifelines like water and energy is on rise. However, an unfortunate water-energy nexus exists as the existing water purification technologies consume excessive amount of power to provide us with potable water. Microbial fuel cells offer an alternate exciting opportunity as they not only cleanse the water but also generate power during the process. If we can take advantage of such dual benefits properly then we can find an effective critical technology for our survivability. However, challenges exist to translate such innovation from lab testbed to market. Due to lengthy biochemical reaction, it is increasingly time exhaustive process. Therefore, nanotechnology can be used in microbial fuel cells from different aspects to enhance their performance, functionalities, simplification in manufacturing and in operation and cost effectiveness. Nanomaterials can be used as anode and cathode electrodes. Nanostructured membranes can also be used for improved filtration. This chapter provides a synoptic picture of advances in the recent years made through nanotechnology, more specifically nanomaterials.

We learnt about using nanoporous structured filters as early as 2007 (Biffinger *et al.* 2007). Proton exchange membranes (PEMs) can restrict the creative diversification in novel cell design to increase the power production. Therefore, Biffinger *et al.* used nanoporous polymer filters (nylon, cellulose, or polycarbonate) in place of PEMs in a miniaturized microbial fuel cell of 2 cm^2 device cross-section area. The cell generated a power density of 16 W/m^3. This power production was comparable to conventional Nafion-117 membranes which are ultra-expensive too. Additionally, the nanoporous membranes isolated the anode chamber to be occupied by natural bacteria. This is an advantage to further use the MFC beyond aquatic sediment environments. This work was most probably the first step to reduce the cost of microbial fuel cells from membrane perspective.

In the same year, Qiao *et al.* used carbon nanotube (CNT) in conjunction with polyaniline (PANI) as anode material for high-power generating microbial fuel cells (Qiao *et al.* 2007). It was observed that the composite anode based cell showed superior performance than that of the cell used PANI only anode. It was asserted that the 20 wt% CNT/PANI composite anode had the highest electrochemical activity with maximum power generation density of 42 mW/m^2 with *Escherichia coli* as the microbial catalyst. This study also showed superior performance than that of other cells using *E. coli* as catalyst but not using CNT based composite anode.

Qiao *et al.* further experimented with a unique nanostructured polyaniline (PANI)/mesoporous TiO_2 composite as anode in *E. coli* microbial fuel cells (MFCs). Various physical metrology techniques like X-ray diffraction, morphology and nitrogen adsorption-desorption studies were carried out to observe a networked nanostructure with uniformly distributed nanopores and high specific surface area. Optimized composite anode with 30 wt% PANI showed excellent bio and electrocatalytic performance and higher power density of 1495 mW/m^2. It is well known that the charge capacity of an electrode for a specific reaction is proportional to the electrode surface. This study showed with increased catalytic reaction area with increased PANI content, the electrical performance also increased until it reached its maximum.

In 2008; Sharma *et al.* reported using carbon nanotube (CNT) based anodes and nanofluids as novel electron mediators in the microbial fuel cells (Sharma *et al.* 2008). The use of nanofluids for microbial fuel cells was the first in its kind. They were prepared by disbanding nanocrytsalline platinum anchored CNTs in water. Performance comparison of this cell with microbial fuel cells with graphite anode and Neutral Red (NR) and Methylene Blue (MB) electron mediators in *E. coli* catalytic environment showed nearly 6 fold increased power density (2470 mW/m^2). This result also showed the capability of nanofluids in effectively channeling the bacterial energy, resulting into excellent power output. Finally, the authors also predicted that using a hybrid mixture of bacteria or electrigens like Shewanella putrefaciens, Geobacter sulfurreducens, Rhodoferax ferrireducens, Pseudomonas aeruginosa etc. in place of *E. coli* may increase the power output further.

In the year 2010; Zou *et al.* successfully used nanomaterials in sun-powered or photosynthetic microbial fuel cells (PMFCs) in a carbon dioxide (CO_2) and organic fuel free environment (Zou *et al.* 2010). Nanostructured electrically conductive polymer polypyrrole showed significant improvement in efficiently collecting electron from photosynthetic biofilm in the cells. Further studies involving cyclic voltammetry and impedance spectroscopy showed extensive advances in electrochemical properties like higher redox current and lower interface electron-transfer resistance. Therefore with a loading density of 3 mg/cm^2, nanostructured fibrillary polypyrrole coated anode showed 450% increased power density. This study opened up opportunity for using conducting polymers with high π-conjugated polymeric chain and high electrical conductivity. With their long chains, they can easily intercalated into cell membranes resulting into direct discharge of electrons to the anode surface.

To scale-up and to initiate commercialization, Liu *et al.* demonstrated enhanced oxygen reduction capacity from cost-effective metal catalysts in an air cathode integrated microbial fuel cell (Liu *et al.* 2010). They used nanostructured manganese oxide (MnO_x) formed by electrochemical deposition to serve as a catalyst for oxygen reduction. Such deposition can be easily used to control the size and morphology of the nanostructures in adaptive fashion. This helped in controlling its catalytic activity and cathodic oxygen reduction performance. Further studies showed MnO_x nanorods had an electrochemical activity towards oxygen reduction reaction *via* a four-electron pathway in a neutral pH solution. They obtained maximum power density of 772.8 mW/m^3 while removing the organics by feeding the cell with an acetate-laden synthetic wastewater. With this work, they demonstrated for the first time a cost-effective facile preparation of cathodic catalyst in a controlled fashion.

As an unusual material, Higgins *et al.* reported in 2011 where they showed Chitosan (CHIT) scaffolds doped with multi-walled carbon nanotubes (MWCNT) as anodic material (Higgins *et al.* 2011). With high resolution imaging they observed direct colonization of CHIT-MWCNT scaffolds with Shewanella oneidensis. They also observed 2.5 vol% glyoxal as optimal cross-linker in context of porosity and structural stability. Further studies showed a maximum power density of 4.75 W/m^3 at a current density of 16 A/m^3 indicating the conjugation of CHIT-MWCNT scaffolds into carbon felt electrodes.

A conformal coating of carbon nanotube on textile fibers formed by Xie *et al.* in 2011 made a macroscale two-fold porous structure to be used as anode (Xie *et al.* 2010). The material stack was highly conductive and biocompatible. The deterministic porosity enabled an open 3D space for efficient substrate transport and internal colonization by a diverse microflora, resulting in a 10-fold-larger anolyte-biofilm-anode interfacial area than the projective surface area of the CNT-textile. The conformal deposition of CNT displayed strong interaction with the microbial biofilm, facilitating electron transfer from exoelectrogens to the CNT-textile anode. An MFC equipped with a CNT-textile anode had a 10-fold-lower charge-transfer resistance and achieves considerably better performance than one equipped with a traditional carbon cloth anode: the maximum power density was 68% higher.

Same year, Lamp *et al.* demonstrated that flame synthesis can provide a highly manufacturable low-cost route for carbon nanostructures (CNSs) on stainless steel with high throughput in an energy-efficient manner (Lamp *et al.* 2011). They applied CNSs to stainless steel with high conductivity for improved anode power densities in MFCs. They achieved record maximum anodic power density of 493 mW/m^2 (based on projected surface area). However, they fall short of explaining an mechanism for increased power density.

S. Chen *et al.* demonstrated a binder-free cathode: nitrogen doped carbon nanofibers on stainless steel mesh (NCNFs/SSM) to show high electrocatalytic activity towards ORR due to excellent properties of the cathode: good flexibility, high conductivity, and connectivity (Chen *et al.* 2012). By using low-cost simple pyrolysis of nitrogen-containing organic matters, they grew NCNFs onto SSM. However, they also did not report on exact mechanism.

In March 2011; Feng *et al.* reported a study on nitrogen-doped carbon nanotubes (NCNTs) as the cathodic catalyst for oxygen reduction in MFCs for efficient and sustainable production of electricity (An *et al.* 2011). They obtained a maximum power density of 1600 ± 50 mW/m^2 which was higher than status-quo expensive platinum (Pt) catalyst (1393 ± 35 mW/m^2). Additionally drop percentage of power densities were higher in NCNTs than that with Pt catalysts by 25 cycles. Feng *et al.* further explained that bamboo-shaped and vertically aligned NCNTs had lower internal resistance and higher cathode potentials yielding a higher electrocatalytic activity for the oxygen reduction reaction (ORR) *via* a four-electron pathway in neutral pH phosphate buffer solution (PBS). They also showed with cyclic voltammograms that NCNTs had better durability for cathodic ORR to drive the overall sustainability.

For the first time, An *et al.* in 2011 tested silver nanoparticles (AgNPs) as a cathodic catalyst for microbial fuel cells to eliminate organic contamination and oxygen depletion (Feng *et al.* 2011). AgNPs integrated cell showed the highest DO concentration (0.8 mg/L) in the cathode area and the highest current (up to 0.12 mA). The authors inferred that the growth of oxygen consuming heterotrophic microbes were inhibited by AgNPs. They also gave rise of DO concentration in the cathode reducing oxygen and ultimately resulted in the highest current output.

N. Malvankar *et al.* reported a highly conductive living, organic material comprising a network of nano-filaments derived from natural amino acids (Malvankar *et al.* 2011). They showed they can engineer metallic functionality into natural, self-renewing, nanostructured materials. They envisioned it might offer possibilities for overcoming barriers associated with coupling abiotic and biotic-materials in microbial fuel cells.

J Ji *et al.* later applied a new vertically aligned integrations strategy with a layer-by-layer assembling technique to have an indium tin oxide (ITO) modified by iron oxide (Fe_2O_3) nanorods and chitosan to be used as the anode in MFC system (Ji *et al.* 2011). This self-assembly process enabled the nanorods to form an integrated free-standing interdigitated or interwoven 3D networked structure. This interesting architecture allowed to have higher quantity of electric charge output with 320% enhancement due to low internal resistance, higher sensitivity and larger active electrode areas for bacterial interactions, comparing to the bare anode.

In the year 2012, U. Karra *et al.* reported the use of activated carbon nanofibers (ACNF) as the anode material to enhance bacterial biofilm growth to improve MFC performance (Karra *et al.* 2013). Further analysis showed that ACNF exhibited better performance over the other commonly used carbon anodes (granular activated carbon (GAC), carbon cloth (CC)). Batch-scale tests demonstrated ACNF and GAC integrated MFCs with 3.5 ± 0.46 W/m^3 and 3.09 ± 0.33 W/m^3, respectively. From contaminant removal efficiency, ACNF integrated MFC showed $85 \pm 4\%$ demonstrating unique advantages of ACNF in terms of biofilm growth and electron transport.

M. Ghasemi *et al.* also independently reported similar result by using self-made carbon nanofiber (CNF)/Nafion and activated carbon nanofiber (ACNF)/Nafion nanocomposite membranes integrated MFCs (Ghasemi *et al.* 2012). They used electrospinning method for production of CNFs due to its simplicity and low-cost. They used nanocomposite membranes to achieve higher power output (57.64 mW/m^2) and coulombic efficiency.

Y-C. Yong *et al.* presumably for the first time, used a novel 3D graphene/PANI structure as the MFC anode (Yong *et al.* 2012). From bacterial biofilm loading and extracellular electron transfer perspectives, it outperformed standard carbon cloth due to its large specific surface area to integrated bacterial biofilms three-dimensionally. This is also a pragmatic approach for largescale MFCs because of its lightweight. Additionally power output can be further enhanced by increasing the thickness. Since 3D graphene electrode is monolithic and has atomically smooth surface, the bacterial membranes will not be disrupted due to penetration of sharp nanoscale features and cytotoxicity will not increase due to update of nanomaterials.

Later M. Hosseini *et al.* reported Titanium (Ti)/nano-Titanium Oxide (TiO_2)/Paladium (Pd) nano-structure cathode electrode in a dual-chambered microbial fuel cell with graphite anode and Flemion cation exchange membrane (Hosseini *et al.* 2012). This new cathode electrode exhibited high endurance to reduced water dissolved oxygen with a maximum output power of 200 mW/m^2. The main advantage of this cathode was its cost-effectiveness compared to Platinum (Pt) doped carbon paper.

In 2013, H. Wang *et al.* demonstrated a different derivative of graphene (Wang *et al.* 2013). They reported a three-dimensional (3D) reduced graphene oxide–nickel (rGO–Ni) foam anode integrated MFC through controlled deposition of rGO sheets onto the nickel foam substrate. They controlled the loading amount of rGO sheets and electrode surface area by the number of rGO loading cycles. In addition to large accessible surface area for microbial colonization and electron mediators, the rGO-Ni offered uniform macro-porous scaffold for effective mass diffusion of the culture medium. This resulted into a significant volumetric power density of 661 mW/m^3. Next they also reported its continued operation in batch-mode for at least a week.

In 2014, S. Shahgaldi *et al.* reported a new composite membrane (polyvinylidene fluoride) PVDF nanofibres synthesized *via* simple, cost effective, highly manufacturable electrospinning to show maximum power density with Nafion 4 membrane, higher than that of Nafion 117 (Shahgaldi *et al.* 2014). Nonetheless, Nafion 2 membrane generated less electricity than all of the other membranes and the Nafion 6 membrane produced approximately the same amount of power as Nafion 117 and Nafion 6. It can be attributed to the fully interconnected PVDF nanofiber bundle.

Y. Wang *et al.* then demonstrated another novel electrode nanomaterial: carbon felt-supported nano-molybdenum carbide (Mo_2C)/carbon nanotubes (CNTs) composite, as platinum-free anode for high-performance MFC (Wang *et al.* 2014). They used microwave-assisted method to synthesize Mo_2C/CNTs from a single source precursor $Mo(CO)_6$. They observed that the carbon felt electrode with 16.7 wt% Mo-Mo_2C/CNTs composite exhibited a comparable electro catalytic activity to that with 20 wt% Platinum (Pt) as anode electro catalyst. This was due to the bifunctional electro catalysis of Mo2C/CNTs for the conversion of organic substrates in to electricity through bacteria. The composite facilitated the formation of biofilm, which was necessary for the electron transfer *via* c-type cytochrome and nanowires. On the other hand, the composite exhibited the electrocatalytic activity towards the oxidation of hydrogen, which was the common metabolite of *E. coli*.

Later that year, Q. Wen *et al.* reported porous nitrogen-doped carbon nanosheet on graphene (PNCN) as an alternative cathode catalyst for ORR in air-cathode MFCs (We *et al.* 2014). Again the pursuit was not restrict for material only but also for low-cost, manufacturable synthetic method for preparing PNCN *via* the carbonization of graphite oxide–polyaniline hybrid (GO–PANI), subsequently followed by KOH activation treatment. Due to its high concentration of nitrogen and high specific surface area, PNCN exhibited an excellent catalytic activity for ORR resulting in a maximum power density of 1159.34 mW/m^2 obtained with PNCN catalyst was higher than that of Pt/C catalyst (858.49 mW/m^2) in a MFC.

R. Amade *et al.* also explored low-cost alternative to Pt, with vertically aligned carbon nanotubes (CNTs) deposited by plasma-enhanced chemical vapor deposition as cathode material in single chamber MFCs (Amade *et al.* 2014). To improve the MFC performance and to lower the ORR, they used manganese oxide (MnO_2) catalyst to electrochemically deposit on the CNTs. It resulted in the output voltage and power density by two orders of magnitude when compared to plain stainless steel electrodes. They also reported that development of microbial communities as biofilm can be significantly limited by substrate metabolic liability. The presence of a labile organic source (acetate) promoted an increase of one order of magnitude in cell performance.

With the growing interest in using CNT, C. Erbay *et al.* reported in January 2015, Through systematic studies on how different lengths, packing densities, and surface conditions of CNTs affect MFC power output, they concluded that long and loosely packed CNTs without any amorphous carbon show the highest power production performance (Erbay *et al.* 2015). They synthesized tentacle-like CNTs directly grown from stainless steel meshes in the radial direction of mesh wires, providing microbes with three-dimensional and large surface areas. As anticipated, CNTs captured microbes like tentacles, resulting in excellent charge transfer characteristics presumably due to pep stacking between the graphitic carbon rings of CNTs and the pili of microbes. The new electrodes had their minimal ohmic loss due to the direct growth of CNTs from stainless steel meshes. They also maintained large enough spaces between CNTs for microbes to be intercalated in between. They also found that the graphitic layer of CNTs with minimum amorphous carbon provided an excellent substrate for charge interaction. The CNT-stainless steel mesh electrode was further tested with wastewater and its long-term performance showed maximum power density of 3360 mW/m^2, 7.4-times higher than that from carbon cloth, being amongst the highest power improvements in nanomaterial-based MFC anodes.

C. Eraby *et al.* also demonstrated highly-porous, light-weight, and inexpensive three-dimensional (3D) sponges consisting of interconnected carbon nanotubes (CNTs) without base materials (Celal Erbay *et al.* 2015). They synthesized it with a facile and scalable one-step chemical vapor deposition process as anode of microbial fuel cells (MFCs). The MFCs generated high power densities of 2150 W/m^3 or 170 W/m^3. This is due to excellent charge transfer between CNTs and microbes. They also presented a cost calculation that material production cost for these CNT sponge would be ~$0.1/gCNT, significantly lower than that of other methods. Additionally, their production ate is also high 3.6 g/h.

M. Ma *et al.* focused on reduction of oxygen reduction reaction (ORR) (Ma *et al.* 2015). They developed a graphitic carbon (GC) based nano silver/iron oxide (AgNPs/Fe_3O_4/GC) composites from waste pomelo skin and used as antibacterial oxygen reduction catalysts for MFCs. AgNPs and Fe_3O_4 were introduced in situ into the composites by one-step carbo-thermal reduction, enhancing their conductivity and catalytic activity. They obtained a maximum power density of 1712 ± 35 mW/m^2 is obtained by AgNPs/Fe3O4/GC washed with sulfuric acid (1 mol L^{-1}) for 1 hour, which declined by 4.12% after 17 cycles. Under catalysis of all AgNP-containing catalysts, ORR proceeded *via* the 4e$^-$ pathway, and no toxic effects to anode microorganisms resulted from inhibiting the cathodic biofilm overgrowth. Most of the AgNPs-containing composites exhibited remarkable power output and coulombic efficiency through lowering proton transfer resistance and air-cathode biofouling.

C. Zhao *et al.* showed flexible and freestanding polypyrrole nanotube membrane can be effective to improve the MFC performance (Zhao *et al.* 2015). They attributed this to larger surface area resulted from the porous structure of the membrane for enhanced interaction with bacteria; the enhanced EET between the polypyrrole membrane anode and the bacterial biofilm due to the large active area for bacterial adhesion and electron acceptance from electron shuttles (riboflavin) in solution; and the increased electron transfer efficiency because of the good electrical conductivity of polypyrrole nanotubes which could be synthesized in large-scale. They also reported its maximum power density of 612 mW/m^2 at a current density of 2.1 A/m^2.

Later last year, S. Zhao *et al.* reported a three-dimensional (3D) graphene aerogel (GA) decorated with platinum nanoparticles (Pt NPs) as an efficient freestanding anode for MFCs. With a continuous 3D macroporous structure, it offered favorable condition for microorganism immobilization and efficient electrolyte transport

(Zhao *et al.* 2015). Homogenous decoration of it with Pt NPs further enhanced extracellular charge transfer between the bacteria and the anode. The MFCs generated a remarkable maximum power density of 1460 mW/m^2.

We investigated potential impact of nanotechnology in enhancing the efficiency of micro-scaled MFCs since 2012. J. E. Mink *et al.* reported integration of vertically multi-walled carbon nanotubes (MWCNTs) in a silicon based micro-sized MFC (Figure 7.1) with complementary metal oxide semiconductor (CMOS) technology to enable reliable batch fabrication (Mink *et al.* 2012). For the first time they introduced Ohmic contact using nickel salicidation for improved output current. We used MWCNTs as they have high surface area-to-volume ratio (66,000/cm before functionalization) to enhance bacterial colonization. Additionally they have ultra-high electrical conductivity (resistance of 2.1 mΩ m for a single nanotube). Both of them allow enhanced transfer of electrons. Finally, MWCNTs are biocompatible, chemically stable, catalytically active and resistive to decomposition. We also performed functionalization of the CNTs with acid treatment (nitric and sulfuric acids) to remove residual metal catalysts and other impurities. They also improved cell adhesion by thickening the MWCNTs in an oxidation process which generated carboxylic groups in the walls and tips of the MWCNTs, formed 3D structures using capillary tensile forces. Another interesting aspect is forest type growth of MWCNTs on nickel seed layer which is analogous to a brush type carbon architecture which allows periodic removal of accumulated sediments from the waste water. Use of nickel as seed layer served two purposes: (i) pragmatic approach for high-volume manufacturing and (ii) low-cost alternate solution which is CMOS foundry compatible compared to prohibited metals like gold (Au). The same nickel was also used to form Ohmic contact: nickel silicide. This provided good electrical contact and avoided the need to make contact with the anode inside the solution chamber. The MFC produced a maximum net power of nearly 500 nW.

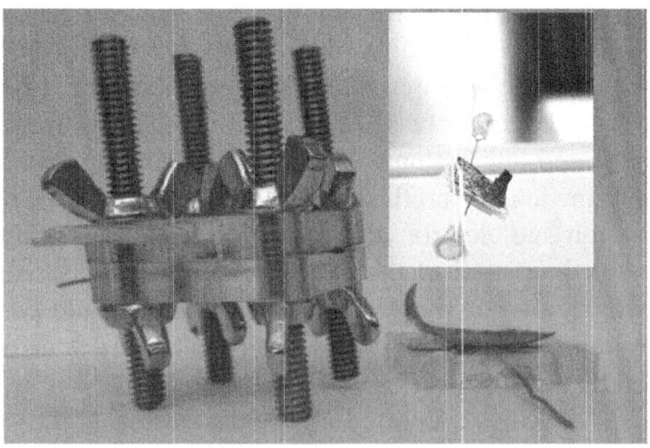

Figure 7.1 Micro-sized microbial fuel cells with full assembly. Inset image shows the carbon cloth glued to polymer casting which is hosting silicon based cell.

To understand the impact of Ohmic *vs.* Schottky contact for solid state MFCs we conducted comprehensive experiments with cobalt (Co) and aluminum (Al) and reported that Ohmic contact with tunneling current mechanism is the most effective way for enhancing output power in MFCs (Mink *et al.* 2013).

Later, we removed the expensive membrane and replaced carbon cloth based cathode with specially designed carbon cloth air cathode in the MWCNTs integrated micro-sized MFCs (Mink *et al.* 2013). In this paper, most importantly we reported 45 days of continuous operation for the MFCs overcoming the long thought hypothesis that air cathode may not be a viable option for MFCs.

With the emergence of graphene and other two dimensional atomic crystal structure materials, we explored graphene based polymeric MFCs with air cathode and obtained nano-scale power output from micro-sized MFCs (Mink *et al.* 2013). We fabricated a 25 μl MFC with graphene as anode materials and air cathode for enabling usage of oxygen present in air. Most importantly for the first time we used human saliva as fuels to make the whole system sustainable and absolutely mobile (Mink *et al.* 2014). Graphene anode integrated MFCs generated 40 times more power than those with traditional carbon cloth as anodes. This study showed us that MFCs can possibly be used for alternate healthcare applications. Presently we are investigating arraying of cells for higher yield (Figure 7.2).

Figure 7.2 Arraying of microbial fuel cells with graphene anode on silicon based platform. Three cells are connected in series for enhanced performance.

In summary, we believe nanotechnology and especially nanomaterials can play critical role in technology translation of MFCs. Carbon based 1D and 2D materials like carbon nanotubes and graphene can thus be instrumental. High conductivity, large interaction areas and biocompatibility are the main attributed sought after in these nanomaterials. Another critical area is to develop low-cost high volume nano-manufacturing process for rapid dissemination of reliable MFCs for wide ranging applications.

7.1 REFERENCES

Amade R., Vila-Costa M., Hussain S., Casamayor E. and Bertran E. (2014). Vertically aligned carbon nanotubes coated with manganese dioxide as cathode material for microbial fuel cells. *Journal of Materials Science*, **50**(3), 1214–1220.

An J., Jeon H., Lee J. and Chang I. (2011). Bifunctional silver nanoparticle cathode in microbial fuel cells for microbial growth inhibition with comparable oxygen reduction reaction activity. *Environmental Science and Technology*, **45**, 5441–5446.

Biffinger J., Ray R., Little B. and Ringeisen B. (2007). Diversifying biological fuel cell designs by use of nanoporous filters. *Environmental Science and Technology*, **41**, 1444–1449.

Celal E., Yang G., Figueiredo P., Sadr R., Yu C. and Han A. (2015). Three-dimensional porous carbon nanotube sponges for high-performance anodes of microbial fuel cells. *Journal of Power Sources*, **298**, 177–183.

Chen S., Chen Y., He G., He S., Schröder U. and Hou H. (2012). Stainless steel mesh supported nitrogen-doped carbon nanofibers for binder-free cathode in microbial fuel cells. *Biosensors and Bioelectronics*, **34**, 282–285.

Erbay C., Pu X., Choi W., Choi M., Ryu Y., Hou H., Lin F., Figueiredo P., Yu C. and Han A. (2015). Control of geometrical properties of carbon nanotube electrodes towards high-performance microbial fuel cell. *Journal of Power Sources*, **280**, 347–354.

Feng L., Yan Y., Chen Y. and Wang L. (2011). Nitrogen-doped carbon nanotubes as efficient and durable metal-free cathodic catalysts for oxygen reduction in microbial fuel cells. *Environmental Science and Technology*, **4**, 1892.

Ghasemi M., Shahgaldi S., Ismail M., Yaakob Z. and Daud W. (2012). New generation of carbon nanocomposite proton exchange membranes in microbial fuel cell systems. *Chemical Engineering Journal*, **184**, 82–89.

Higgins S., Foerster D., Cheung A., Lau C., Bretschger O., Minteer S., Nealson K., Atanassov P. and Cooney M. (2011). Fabrication of macroporous chitosan scaffolds doped with carbon nanotubes and their characterization in microbial fuel cell operation. *Enzyme and Microbial Technology*, **48**, 458–465.

Hosseini M. and Ahadzadeh I. (2012). A dual-chambered microbial fuel cell with Ti/nano-TiO_2/Pd nano-structure cathode. *Journal of Power Sources*, **220**, 292–297.

Ji J., Jia Y., Wu W., Bai L., Ge L. and Gu Z. (2011). A layer-by-layer self-assembled Fe_2 O_3 nanorod-based composite multilayer film on ITO anode in microbial fuel cell. *Colloids and Surfaces A: Physicochem. Eng. Aspects*, **390**, 56–61.

Karra U., Manickam S., McCutcheon J., Patel N. and Li B. (2013). Power generation and organics removal from wastewaterusing activated carbon nanofiber (ACNF) microbial fuel cells (MFCs). *International Journal of Hydrogen Energy*, **38**, 1588–1597.

Lamp J., Guest J., Naha S., Radavich K., Love N., Ellis M. and Puri I. (2011). Flame synthesis of carbon nanostructures on stainless steel anodes for use in microbial fuel cells. *Journal of Power Sources*, **196**, 5829–5834.

Liu X., Sun X., Huang Y., Sheng G., Zhou K., Zeng R., Dong F., Wang S., Xu A., Tong Z. and Yu H. (2010). Nano-structured manganese oxide as a cathodic catalyst for enhanced oxygen reduction in a microbial fuel cell fed with a synthetic wastewater. *Water Research*, **44**, 5298–5305.

Ma M., You S., Gong X., Dai Y., Zou J. and Fu H. (2015). Silver/iron oxide/graphitic carbon composites as bacteriostatic catalysts for enhancing oxygen reduction in microbial fuel cells. *Journal of Power Sources*, **283**, 74–83.

Malvankar N., Vargas M., Nevin K., Franks A., Leang C., Kim B., Inoue K., Mester T., Covalla S., Johnson J., Rotello V., Tuominen M. and Lovley D. (2011). Tunable metallic-like conductivity in microbial nanowire networks. *Nature Nanotechnology*, **6**, 573–579.

Mink J., Hussain M. (2013). Sustainable design of high performance micro-sized microbial fuel cell with carbon nanotube anode and air cathode. *ACS Nano*, **7**(8), 6921–6927.

Mink J., Rojas J., Logan B. and Hussain M. (2012). Vertically grown multi-walled carbon nanotube anode and nickel silicide integrated high performance micro-sized (1.25 μL) microbial fuel cell. *Nano Letters*, **12**(2), 791–795.

Mink J., Rojas J. and Hussain M. (2013a). Role of contact engineering in micro-sized microbial fuel cell. *Physica Status Solidi A*, **211**(3), 550–554.

Mink J., Qaisi R. and Hussain M. (2013b). Graphene based flexible micro-sized microbial fuel cell. *Energy Technology*, **1**(11), 648–652.

Mink J., Qaisi R., Logan B., Hussain M. (2014). Energy harvesting from saliva in micro-sized microbial fuel cells. *NPG Asia Materials*, **6**, e89.

Qiao Y., Bao S., Ming Li C., Cui X., Lu Z. and Guo J. (2007a). Nanostructured polyaniline/titanium dioxide composite anode for microbial fuel cells. *American Chemical Society*, 13–119.

Qiao Y., Li C. Bao S. and Bao Q. (2007b). Carbon nanotube/polyaniline composite as anode material for microbial fuel cells. *Journal of Power Sources*, **170**, 79–84.

Shahgaldı S., Ghasemi M., Daud W., Yaakob Z., Sedighi M., Alam J. and Ismail A. (2014). Performance enhancement of microbial fuel cell by PVDF/Nafion nanofibre composite proton exchange membrane. *Fuel Processing Technology*, **124**, 290–295.

Sharma T., Reddy A., Chandra T. and Ramaprabhu S. (2008). Development of carbon nanotubes and nanofluids basedmicrobial fuel cell. *International Journal of Hydrogen Energy*, **3**, 6749–6754.

Wang H., Wang G., Ling Y., Qian F., Song Y., Lu X., Chen S., Tong Y. and Li Y. (2013). High power density microbial fuel cell with Flexible 3D grapheme. *The Royal Society of Chemistry (Nanoscale)*, 10283–10290.

Wang Y., Li B., Cui D., Xiang X. and Li W. (2014). Nano-molybdenum carbide/carbon nanotubes composite as bifunctional anode catalyst for high-performance *Escherichia coli*-based microbial fuel cell. *Biosensors and Bioelectronics*, **51**, 349–355.

We Q., Wang S., Yan J., Cong L., Chen Y. and Xi H. (2014). Porous nitrogen-doped carbon nanosheet on graphene as metal-free catalyst for oxygen reduction reaction in air cathode microbial fuel cells. *Bioelectrochemistry*, **95**, 23–28.

Xie X., Hu L., Pasta M., Wells G., Kong D., Criddle C. and Cui Y. (2010). Three-dimensional carbon nanotube-textile anode for high-performance microbial fuel cells. *American Chemical Society*, **11**(1), 291–296.

Yong Y., Dong X., Chan-Park M., Song H. and Chen P. (2012). Macroporous and monolithic anode based on polyaniline hybridized three-dimensional graphene for high-performance microbial fuel cells. *American Chemical Society*, **6**(3), 2394–2400.

Zhao C., Wu J., Kjelleberg S., Loo J. and Zhang Q. (2015). Employing a flexible and low-cost polypyrrole nanotube membrane as an anode to enhance current generation in microbial fuel cells. *Small Journal*, **11**(28), 3440–3443.

Zhao S., Li Y., Yin H., Liu Z., Luan E., Zhao F., Tang Z. and Liu S. (2015). Three-dimensional graphene/Pt nanoparticle composites as freestanding anode for enhancing performance of microbial fuel cells. *Science Advances*, **1**(10), e1500372.

Zou Y., Pisciotta J. and Baskakov I. (2010). Nanostructured polypyrrole-coated anode for sun-powered microbial fuel cells. *Bioelectrochemistry*, **79**(1), 50–56.